IRRATIONAL
NUMBERS

By
IVAN NIVEN

THE
CARUS MATHEMATICAL MONOGRAPHS

Published by

THE MATHEMATICAL ASSOCIATION OF AMERICA

———

THE CARUS MATHEMATICAL MONOGRAPHS are an expression of the desire of Mrs. Mary Hegeler Carus, and of her son, Dr. Edward H. Carus, to contribute to the dissemination of mathematical knowledge by making accessible at nominal cost a series of expository presentations of the best thoughts and keenest researches in pure and applied mathematics. The publication of the first four of these monographs was made possible by a notable gift to the Mathematical Association of America by Mrs. Carus as sole trustee of the Edward C. Hegeler Trust Fund. The sales from these have resulted in the Carus Monograph Fund, and the Mathematical Association has used this as a revolving book fund to publish the succeeding monographs.

The expositions of mathematical subjects which the monographs contain are set forth in a manner comprehensible not only to teachers and students specializing in mathematics, but also to scientific workers in other fields, and especially to the wide circle of thoughtful people who, having a moderate acquaintance with elementary mathematics, wish to extend their knowledge without prolonged and critical study of the mathematical journals and treatises. The scope of this series includes also historical and biographical monographs.

The following monographs have been published

The Carus Mathematical Monographs

NUMBER ELEVEN

IRRATIONAL
NUMBERS

By

IVAN NIVEN

Professor of Mathematics
University of Oregon

Published by
THE MATHEMATICAL ASSOCIATION OF AMERICA

Distributed by
JOHN WILEY AND SONS, INC.

512.7

512

15430

Composed and Printed
by
Quinn & Boden Company, Inc.
Rahway, New Jersey
1956

PREFACE

This monograph is intended as an exposition of some central results on irrational numbers, and is not aimed at providing an exhaustive treatment of the problems with which it deals. The term "irrational numbers," a usage inherited from ancient Greece which is not too felicitous in view of the everyday meaning of the word "irrational," is employed in the title in a generic sense to include such related categories as transcendental and normal numbers.

The entire subject of irrational numbers cannot of course be encompassed in a single volume. In the selection of material the main emphasis has been on those aspects of irrational numbers commonly associated with number theory and Diophantine approximations. The topological facets of the subject are not included, although the introductory part of Chapter I has a sketch of some of the simplest set-theoretic properties of the irrationals as a part of the continuum. The axiomatic basis for irrational numbers, proceeding say from the Peano postulates for the natural numbers to the construction of the real numbers, is purposely omitted, because in the first place the aim is not in the direction of the foundations of mathematics, and in the second place there are excellent treatments of this topic readily available.

The customary organization of a book with related subjects grouped together has been modified in part by consideration of the degree of difficulty of the topics, proceeding from the easiest to the most difficult. For example, almost all the theorems on irrational numbers in Chapter II are implied by the stronger results of Chapter IX, but, whereas Chapter II requires only calculus and the barest rudiments of number theory for understanding, Chapter IX presupposes some basic results on algebraic numbers and complex functions. The first seven chapters are distinctly easier reading than the last three, with fewer prerequisite results needed and less mathematical maturity required of the reader. The chapters are for the most part independent of one another and so can be read separately; the major exception to this statement is the use in Chapter VI of some results from Chapter V.

The only knowledge required of the reader beyond quite elementary mathematics is some algebraic number theory in Chapters III, IX and X, and some function theory in Chapters VI, VIII, IX and X. Most of the results needed are well-known theorems, central to the mainstream of mathematics, and complete references are given to standard works. In those few instances where the prerequisite material is at all special, it has been included in the text.

The books by Hardy and Wright, Koksma, Perron, and Siegel listed on page 157 have been very helpful, and I have made free use of these excellent sources. Further source material is listed in the notes at the ends of the chapters. These references, along with the remarks in the Notes, may be taken or left alone at the reader's choice. Some further results beyond the scope of this book are also listed in the Notes; however, as any expert in the subject will readily see, I have not attempted to be either systematic or complete about this. For the convenience of the reader there is appended a list of notation and a glossary on pages 151 to 156.

*   *   *

Substantial improvements in the book have resulted from discussion of many points with my colleagues at the University of Oregon, and from bibliographic suggestions by Professor C. D. Olds. I am also indebted to the Editorial Committee of the Carus Monographs for help in removing several errors and obscurities. But especially I wish to acknowledge my indebtedness to Professor H. S. Zuckerman who has been actively interested in this project from the start. Discussions with him during the early stages influenced markedly the final versions of Chapters I and V. In addition he has read the manuscript very thoroughly and critically. However I did not invariably follow the suggestions of these friendly critics; so the responsibility for the shortcomings of the monograph is entirely mine.

IVAN NIVEN

University of Oregon
July 1956

CONTENTS

CHAPTER 1

RATIONALS AND IRRATIONALS

1. The preponderance of irrationals. Our general intention in this book is to characterize, classify, and exhibit irrational numbers in various ways, not only in the framework of real numbers but also in the larger setting of complex numbers. We do not examine the axiomatic foundations of our subject, preferring simply to take for granted the following basic classification. A rational number is one that can be put in the form h/k, where h and k are integers with $k \neq 0$. Real numbers like $\sqrt{2}$ which are not rational are said to be irrational. †

The first three sections of this chapter are devoted to some observations on the rational and irrational numbers regarded as point sets on the real line. This set-theoretic analysis is rather cursory, and is not typical of the general line of thought of this monograph, which is more number-theoretic in character.

We begin by drawing attention to the overwhelming preponderance of irrationals over rationals. This cannot be established by a simple count, since both the rational numbers and the irrational numbers constitute infinite sets. To make a comparison of these two sets, we think

† In addition to definitions and explanation of terminology given in the text, there is a Glossary and List of Notation at the end of the book.

of the real numbers as points on a line in the customary
fashion in coordinate geometry. Any interval on this line,
for example the interval from 3 to 5, or all x satisfying
$3 \leq x \leq 5$, covers all rational and irrational points in the
interval. What we will establish is that we can create a
set of intervals which cover all the rational points on the
real line, and yet the total of the lengths of the intervals
is arbitrarily small. The irrational numbers do not pos-
sess this property.

A set S of real numbers is said to have *measure zero* if it
is possible to cover the points of S with a set of intervals
of arbitrarily small total length. To give a simple example, the set of positive integers has measure zero as can
be seen as follows. Enclose the integer 1 in the interval
$(1 - \epsilon/2, 1+\epsilon/2)$, the integer 2 in the interval $(2 - \epsilon/4,
2 + \epsilon/4)$, $\cdots$, in general the integer n in the interval
$(n - \epsilon/2^n, n + \epsilon/2^n)$. Hence the positive integers are
covered by intervals of total length $\epsilon + \epsilon/2 + \epsilon/4 + \cdots
= 2\epsilon$, which can be made arbitrarily small.

Next, we say that *almost all* real numbers in an interval
have a certain property P if the set of real numbers (or
points) lacking the property P has measure zero. For example by the previous paragraph we may say that almost
all positive real numbers are non-integers. This is a special case of the following result.

THEOREM 1.1. *Almost all real numbers are irrational.*

Proof. First we prove that the positive rational numbers in the unit interval $(0, 1)$ have measure zero. These
numbers are the rationals h/k with $1 \leq h \leq k$, and we enclose each such rational in an interval

(1.1) $(h/k - \epsilon/k^3, \quad h/k + \epsilon/k^3)$.

We could avoid unnecessary duplication of the rationals
by imposing the restriction that h and k be relatively

prime, but the present proof is correct with or without this restriction. These intervals (1.1) have total length which is

$$\leqq \sum_{k=1}^{\infty} \sum_{h=1}^{k} \frac{2\epsilon}{k^3} = \sum_{k=1}^{\infty} \frac{2\epsilon}{k^2} < 2\epsilon + \sum_{k=2}^{\infty} \frac{2\epsilon}{k(k-1)}$$

$$= 2\epsilon + \sum_{k=2}^{\infty} 2\epsilon \left(\frac{1}{k-1} - \frac{1}{k} \right) = 4\epsilon,$$

which can be made arbitrarily small by suitable choice of ϵ. (This argument can be shortened by using $\Sigma k^{-2} = \pi^2/6$.)

Next we extend this covering procedure to all positive rationals by observing that the reciprocal of any $h/k \leqq 1$ is $k/h \geqq 1$ which can be enclosed in an interval

$$(k/h - \epsilon/k^3, \quad k/h + \epsilon/k^3).$$

The estimate of the total lengths of these intervals is identical with that above, and so can be made arbitrarily small. The proof of the theorem is completed by the obvious extension to the negative rational numbers and zero.

It is interesting to note a specific irrational number not covered by the intervals (1.1) for suitable ϵ. Such a number is $\sqrt{2}/2$, as can be seen in the following way. First, the irrationality of $\sqrt{2}$ implies that, for any integers h and k, $| k^2 - 2h^2 | \geqq 1$. Hence for any positive rational $h/k \leqq 1$ we have

$$\left| \frac{\sqrt{2}}{2} - \frac{h}{k} \right| = \left| \frac{k^2 - 2h^2}{2k^2 \left(\dfrac{\sqrt{2}}{2} + \dfrac{h}{k} \right)} \right| \geqq \frac{1}{2k^2 \left(\dfrac{\sqrt{2}}{2} + 1 \right)}$$

$$> \frac{1}{4k^2} \geqq \frac{1}{4k^3}.$$

Thus $\sqrt{2}/2$, thought of as a point on the real axis, has distance from the point h/k greater than $1/4k^3$, and so

$\sqrt{2}/2$ is not in any interval (1.1) provided we choose ϵ suitably small, say $\epsilon = 1/4$.

2. Countability. A set is said to be *countable* (or *denumerable*) if it can be put in one-to-one correspondence with the positive integers $1, 2, 3, \cdots$.

THEOREM 1.2. *The rational numbers are countable.*

Proof. First we give a rule for ordering the positive rational numbers h/k, assuming that $(h, k) = 1$ to avoid duplicates. Say that h_1/k_1 precedes h_2/k_2 in case either of the following conditions (*a*) or (*b*) holds:

(*a*) $$h_1 + k_1 < h_2 + k_2;$$

(*b*) $$h_1 + k_1 = h_2 + k_2 \quad \text{and} \quad h_1 < h_2.$$

This principle gives us an ordering of the positive rationals beginning thus:

$$\frac{1}{1}, \frac{1}{2}, \frac{2}{1}, \frac{1}{3}, \frac{3}{1}, \frac{1}{4}, \frac{2}{3}, \frac{3}{2}, \frac{4}{1}, \frac{1}{5}, \frac{5}{1}, \cdots.$$

This sequence, which includes every positive rational, can be put into one-to-one correspondence with the positive integers. The extension to all rationals can be obtained by inserting the negatives and zero in some such fashion as

$$0, \frac{1}{1}, -\frac{1}{1}, \frac{1}{2}, -\frac{1}{2}, \frac{2}{1}, -\frac{2}{1}, \frac{1}{3}, -\frac{1}{3}, \cdots.$$

THEOREM 1.3. *Any countable set S of real numbers, say $a_1, a_2, a_3, \cdots$, has measure zero.*

Proof. The argument used on the positive integers in § 1 can be employed. That is, enclose a_n in the interval $(a_n - \epsilon/2^n, a_n + \epsilon/2^n)$ for $n = 1, 2, 3, \cdots$. Thus the whole set S is enclosed in intervals of length $\leq 2\epsilon$, and this is arbitrarily small.

It may be noted that Theorem 1.1 is implied by Theorems 1.2 and 1.3.

THEOREM 1.4. *The irrational numbers are not countable. A fortiori the real numbers are not countable.*

Proof. Suppose the irrational numbers were countable, say γ_1, γ_2, γ_3, $\cdots$. But the rationals are countable by Theorem 1.2, say r_1, r_2, r_3, $\cdots$. Thus we could intersperse these two sequences to get a single sequence r_1, γ_1, r_2, γ_2, r_3, γ_3, $\cdots$. Thus the reals would be countable and so would have measure zero by Theorem 1.3. But this would say that the whole real line can be covered by a set of intervals of arbitrarily small total length, a contradiction from which the theorem follows. (The proposition that the whole real line *cannot* be covered by a set of intervals of arbitrarily small total length is in a sense intuitively clear; a rigorous proof, which we do not wish to go into here, involves a more detailed study of the topology of the real line.)

3. Dense sets. A set S of real numbers is said to be *everywhere dense* in an interval if, given any two real numbers α and β in the interval, say with $\alpha < \beta$, there is a number s in S such that $\alpha < s < \beta$.

THEOREM 1.5. *The rationals are everywhere dense; the irrationals are everywhere dense.*

Proof. Let α and β be any two real numbers, with $\alpha < \beta$. By the Archimedean property (cf. glossary) there is a positive integer n such that $n(\beta - \alpha) > 1$ or $\beta - \alpha > 1/n$. Choose the integer m to satisfy $m < n\beta \leqq m + 1$. Thus we have

$$\alpha < \beta - \frac{1}{n} \leqq \frac{m+1}{n} - \frac{1}{n} = \frac{m}{n} \quad \text{and} \quad \frac{m}{n} < \beta,$$

so that m/n is a rational number between α and β.

To obtain an irrational number between α and β, use the Archimedean property again to select a positive integer k so that $k(\beta - m/n) > \sqrt{2}$. Hence we can write

$$\beta > \frac{m}{n} + \frac{\sqrt{2}}{k} > \frac{m}{n} > \alpha,$$

so that $m/n + \sqrt{2}/k$ is an irrational number lying between α and β.

4. Decimal expansions. We shall employ constantly the square-bracket notation, $[\alpha]$, to denote the greatest integer not exceeding the real number α. In other words, $[\alpha]$ is the unique integer m satisfying the inequality $m \leq \alpha < m + 1$. In some circumstances it is convenient to have the least integer not less than α: that is, the unique integer n satisfying $n - 1 < \alpha \leq n$. This can be obtained with the square-bracket notation; thus $n = -[-\alpha]$. By the *fractional part* of any real number α is meant the difference $\alpha - [\alpha]$, which will be denoted by (α).

It is well known that the fractional part of any real number α can be represented as an infinite decimal expansion to base 10; thus

$$(\alpha) = \alpha - [\alpha] = .a_1 a_2 a_3 \cdots.$$

This decimal expansion is unique except that a so-called terminating decimal (one with an infinite succession of zeros) can be expressed with an infinite succession of nines, as for example

$$0.42 = 0.42000\cdots = 0.41999\cdots.$$

The representation of a real number by an infinite decimal has been generalized by Cantor in the following way.

THEOREM 1.6. *Let $a_1, a_2, a_3, \cdots$ be a sequence of positive integers, all greater than 1. Then any real number α is uniquely expressible in the form*

$$(1.2) \qquad \alpha = c_0 + \sum_{i=1}^{\infty} \frac{c_i}{a_1 a_2 \cdots a_i},$$

with integers c_i satisfying the inequalities $0 \leqq c_i \leqq a_i - 1$ for all $i \geqq 1$, and $c_i < a_i - 1$ for infinitely many i.

Proof. It will be convenient to have the identity

$$(1.3) \qquad \sum_{i=1}^{\infty} \frac{a_{n+i} - 1}{a_{n+1} a_{n+2} \cdots a_{n+i}} = 1,$$

which is readily established by the following observation. The sum of the first k terms of the series has the value

$$1 - \frac{1}{a_{n+1} a_{n+2} \cdots a_{n+k}},$$

which tends to 1 as k tends to infinity.

We define the integers $c_0, c_1, c_2, \cdots$ and also a sequence of real numbers $\alpha_1, \alpha_2, \alpha_3, \cdots$ by the equations

$$c_0 = [\alpha], \qquad \alpha_1 = \alpha - c_0$$

$$(1.4)$$

$$c_i = [a_i \alpha_i], \qquad \alpha_{i+1} = a_i \alpha_i - c_i, \qquad i = 1, 2, 3, \cdots.$$

Thus α_i is the fractional part of $a_{i-1} \alpha_{i-1}$, so that

$$(1.5) \qquad 0 \leqq \alpha_i < 1, \qquad i = 1, 2, 3, \cdots.$$

Multiplying by a_i, we get $0 \leqq a_i \alpha_i < a_i$, and, since c_i is the integral part of $a_i \alpha_i$, we conclude that $0 \leqq c_i \leqq a_i - 1$, as required by the theorem.

Next we use equations (1.4) and mathematical induction to observe that

(1.6)

$$\alpha = c_0 + \alpha_1 = c_0 + \frac{c_1}{a_1} + \frac{\alpha_2}{a_1} = c_0 + \frac{c_1}{a_1} + \frac{c_2}{a_1 a_2} + \frac{\alpha_3}{a_1 a_2}$$

$$= \cdots = c_0 + \sum_{i=1}^{n} \frac{c_i}{a_1 a_2 \cdots a_i} + \frac{\alpha_{n+1}}{a_1 a_2 \cdots a_n}.$$

We define x_n as

$$x_n = c_0 + \sum_{i=1}^{n} \frac{c_i}{a_1 a_2 \cdots a_i},$$

so that by (1.5), (1.6), and the hypothesis $a_i \geqq 2$ we have

$$0 \leqq \alpha - x_n = \frac{\alpha_{n+1}}{a_1 a_2 \cdots a_n} < \frac{1}{2^n}.$$

Thus the sequence $\alpha - x_n$ tends to zero as n tends to infinity, and this implies (1.2).

Next we establish that $c_i < a_i - 1$ for infinitely many i. On the contrary, suppose that there is a fixed integer n so that $c_i = a_i - 1$ for all $i > n$. Then we use (1.2) and (1.3) to obtain

$$\alpha = c_0 + \sum_{i=1}^{n} \frac{c_i}{a_1 a_2 \cdots a_i} + \sum_{i=n+1}^{\infty} \frac{a_i - 1}{a_1 a_2 \cdots a_i}$$

$$= c_0 + \sum_{i=1}^{n} \frac{c_i}{a_1 a_2 \cdots a_i}$$

$$+ \frac{1}{a_1 a_2 \cdots a_n} \sum_{i=1}^{\infty} \frac{a_{n+i} - 1}{a_{n+1} a_{n+2} \cdots a_{n+i}}$$

$$= c_0 + \sum_{i=1}^{n} \frac{c_i}{a_1 a_2 \cdots a_i} + \frac{1}{a_1 a_2 \cdots a_n}$$

A comparison of this with (1.6) yields $\alpha_{n+1} = 1$, contrary to (1.5).

Finally we establish the uniqueness of the representation (1.2). Suppose that

$$\alpha = b_0 + \sum_{i=1}^{\infty} \frac{b_i}{a_1 a_2 \cdots a_i},$$

where the integers b_i satisfy the same conditions as do the c_i. We must prove that $b_i = c_i$ for every i. The identity (1.3) implies that

$$\sum_{i=1}^{\infty} \frac{b_i}{a_1 a_2 \cdots a_i} < 1,$$

because of the condition that $b_i < a_i - 1$ for infinitely many i. Hence we see that b_0 is the integral part of α; thus $b_0 = [\alpha]$. Also $c_0 = [\alpha]$ by (1.4), and so $b_0 = c_0$. Next suppose that for some $n \geqq 1$ the pair b_n and c_n are unequal. There is no loss of generality in assuming that n is the smallest integer with this property, and that $c_n > b_n$, so that $c_n - b_n \geqq 1$. Hence we have

$$\sum_{i=n}^{\infty} \frac{c_i}{a_1 a_2 \cdots a_i} = \sum_{i=n}^{\infty} \frac{b_i}{a_1 a_2 \cdots a_i},$$

and, since these series are absolutely convergent, we may rearrange terms to get

$$\sum_{i=n+1}^{\infty} \frac{b_i - c_i}{a_1 a_2 \cdots a_i} = \frac{c_n - b_n}{a_1 a_2 \cdots a_n} \geqq \frac{1}{a_1 a_2 \cdots a_n}.$$

We obtain a quite different inequality on the magnitude of this infinite sum by again using the fact that $b_i < a_i - 1$ for infinitely many i, so that $b_i - c_i < a_i - 1$ for infinitely many i. We employ this and (1.3) to write

$$\sum_{i=n+1}^{\infty} \frac{b_i - c_i}{a_1 a_2 \cdots a_i} < \sum_{i=n+1}^{\infty} \frac{a_i - 1}{a_1 a_2 \cdots a_i}$$

$$= \frac{1}{a_1 a_2 \cdots a_n} \sum_{i=1}^{\infty} \frac{a_{n+i} - 1}{a_{n+1} a_{n+2} \cdots a_{n+i}}$$

$$= \frac{1}{a_1 a_2 \cdots a_n}.$$

Thus we have a contradiction, and the proof of the theorem is complete.

We remarked that Theorem 1.6 is a generalization of the ordinary decimal expansion of a real number α. This can be seen by taking all the integers $a_i = 10$. Thus, if α is positive, the equation (1.2) gives the decimal representation

$$(1.7) \qquad \alpha = c_0 + \sum_{i=1}^{\infty} \frac{c_i}{10^i} = c_0.c_1 c_2 c_3 \cdots.$$

For negative α we get the decimal expansion by first getting $-\alpha$ in the form (1.7) and then changing all signs. It may be noted that any real number α with an ambiguous decimal representation, such as $\alpha = 0.42$, is given by (1.7) in the form with an infinite succession of zeros, thus $\alpha = 0.42 = 0.42000\cdots$. The reason for this is that an infinite succession of nines is ruled out by the condition of the theorem that $c_i < a_i - 1$ for infinitely many i, a condition which in the present special case takes the form $c_i < 9$ for infinitely many i.

We now give conditions which guarantee that the number α represented by (1.2) is irrational.

THEOREM 1.7. *Let the integers a_i be as described in the preceding theorem, and let the integers c_i satisfy the inequalities of that result. Furthermore, assume that an infinite number of the c_i are positive, and that each prime number* *divides infinitely many of the a_i. Then α in (1.2) is irrational.*

Proof. Suppose on the contrary that α is rational; say $\alpha = h/k$. Then by the last hypothesis of the theorem we can choose an integer n sufficiently large so that k is a divisor of the product $a_1 a_2 \cdots a_n$. Then, replacing α in (1.2) by h/k, we multiply by $a_1 a_2 \cdots a_n$ and rearrange terms to get

$$\frac{a_1 a_2 \cdots a_n (h - c_0 k)}{k} - \sum_{i=1}^{n} \frac{c_i a_1 a_2 \cdots a_n}{a_1 a_2 \cdots a_i}$$
$$= \sum_{i=1}^{\infty} \frac{c_{n+i}}{a_{n+1} a_{n+2} \cdots a_{n+i}}.$$

The left side is an integer, and we get a contradiction by establishing that the series on the right side converges to a value between 0 and 1. The series on the right side is positive by the hypotheses, and it is strictly less than

$$\sum_{i=1}^{\infty} \frac{a_{n+i} - 1}{a_{n+1} a_{n+2} \cdots a_{n+i}},$$

which has the value 1 by (1.3).

COROLLARY 1.8.   *e is irrational.*

Proof. The standard infinite series expansion,

$$e = 1 + \frac{1}{1!} + \frac{1}{2!} + \frac{1}{3!} + \cdots,$$

satisfies the hypotheses of Theorem 1.7.

Alternatively, we may avoid the use of Theorem 1.7 and give a direct proof of the irrationality of e as follows. Assuming that $e = h/k$ with k positive, we observe that

$$k! \left(e - 1 - \frac{1}{1!} - \frac{1}{2!} - \frac{1}{3!} - \cdots - \frac{1}{k!} \right)$$

is an integer. But this becomes, when e is replaced by its series expansion,

$$\frac{1}{(k+1)} + \frac{1}{(k+1)(k+2)} + \frac{1}{(k+1)(k+2)(k+3)} + \cdots .$$

A contradiction arises because we can show that this series converges to a positive value less than one, and so cannot be an integer. For the series is strictly less than

$$\frac{1}{(k+1)} + \frac{1}{(k+1)^2} + \frac{1}{(k+1)^3} + \cdots = \frac{1}{k} \leqq 1.$$

This proof may be extended, albeit with some slight difficulty, to establish that e is not a quadratic irrational; i.e., that $ae^2 + be + c = 0$ is impossible non-trivially in integers a, b, c. The trick at the start of the proof is to write the equation as $ae + ce^{-1} = -b$, and to use the series expansion for e^{-1} as well as that for e. We shall not pursue this, however, because in the next chapter we prove more about e in Theorems 2.10 and 2.12 with deeper methods. In particular, Theorem 2.12 is the generalization from the quadratic to the general case of the above assertion about the impossibility of $ae^2 + be + c = 0$. Theorem 2.12, in turn, will be generalized in another way in Chapter 9, Theorem 9.1.

We now return to the topic of decimal expansions, which as we have observed are a special case of Theorem 1.6. The decimal $.a_1a_2a_3 \cdots$ is said to be *periodic* if there exist positive integers r and s such that $a_n = a_{n+r}$ for all $n > s$.

THEOREM 1.9. *A real number α has a periodic decimal expansion if and only if α is rational.*

Proof. Inasmuch as the integral part $[\alpha]$ of α has no relevance here, we may assume that $[\alpha] = 0$, or that $0 \leqq \alpha < 1$. First, if α is a periodic decimal, say

$$\alpha = .a_1 a_2 \cdots a_s b_1 b_2 \cdots b_r b_1 b_2 \cdots b_r \cdots,$$

then the infinite decimal $.b_1 b_2 \cdots b_r b_1 b_2 \cdots b_r \cdots$ is the fractional part of both $10^s \alpha$ and $10^{s+r} \alpha$, so that

$$(1.8) \qquad 10^s \alpha - [10^s \alpha] = 10^{s+r} \alpha - [10^{s+r} \alpha].$$

Thus we can write

$$\alpha = \frac{[10^{s+r} \alpha] - [10^s \alpha]}{10^{s+r} - 10^s},$$

which exhibits α as the quotient of two integers.

Conversely, we assume that α is rational: say $\alpha = h/k$ with $k > 0$. The fractional parts of

$$(1.9) \qquad \alpha, \ 10\alpha, \ 10^2 \alpha, \ 10^3 \alpha, \ \cdots$$

are values in the finite set

$$0, \frac{1}{k}, \frac{2}{k}, \ \ldots, \frac{k-1}{k}.$$

Hence we can find two members, say $10^s \alpha$ and $10^{s+r} \alpha$, of the infinite sequence (1.9) having the same fractional parts, and this gives the equation (1.8). Define the positive integer n by the equation

$$n = [10^{s+r} \alpha] - [10^s \alpha],$$

and then we have from (1.8) that

$$10^s \alpha = n/(10^r - 1).$$

Define $n_1 = [n/(10^r - 1)]$, so that

$$10^s \alpha = n/(10^r - 1) = n_1 + n_2/(10^r - 1),$$

where the integer n_2 satisfies $0 \leqq n_2 < 10^r - 1$. This equation can be rewritten in the form

$$10^s \alpha - n_1 = \frac{n_2}{10^r} + \frac{n_2}{10^{2r}} + \frac{n_2}{10^{3r}} + \cdots,$$

and the infinite sum on the right side is seen to be a periodic decimal because of the inequality satisfied by the integer n_2, whence

$$10^s \alpha - n_1 = .b_1 b_2 \cdots b_r b_1 b_2 \cdots b_r \cdots.$$

By adding n_1 to both sides, and then multiplying by 10^{-s}, we establish α as a periodic decimal.

Notes on Chapter 1

§ 1. The analytic foundation of the theory of real numbers, i.e., the construction of the real numbers from the rational numbers by such methods as Dedekind cuts or Cauchy sequences, is not treated here. Excellent formulations of this can be found in Chapter III of Birkhoff and MacLane, † *Survey of Modern Algebra*, or Chapters III and IV of E. Landau, *Foundations of Analysis*.

The word "incommensurable" is frequently found in the earlier literature. Two line segments, or two numbers, are incommensurable in case their ratio is irrational. The origins of these words and concepts are delineated in O. Neugebauer, *The Exact Sciences in Antiquity*, Princeton (1952); George Sarton, *A History of Science*, Harvard (1952); and Morris Kline, *Mathematics in Western Culture*, Oxford (1953).

§ 2. The word "countable" is frequently used in a broader sense to include also the finite sets; cf. R. L. Wilder, *Introduction to the Foundations of Mathematics*, p. 84.

§ 4. Theorem 1.7 is due to G. Cantor, *Zeit. Math. Phys.*, **14** (1869), 121–128; Collected Papers, Berlin (1932), pp. 35–42. Various extensions of this result have been made; cf. M. R. Spiegel, *Amer. Math. Monthly*, **60** (1953), 27–28; A. Oppenheim, *Amer. Math. Monthly*, **61** (1954), 235–241; P. H. Diananda and A. Oppenheim, *Amer. Math. Monthly*, **62** (1955), 222–225.

† Wherever in the Notes the reference to a book is incomplete as to publisher, edition, etc., complete information is given in the list of Reference Books on p. 157.

CHAPTER 2

SIMPLE IRRATIONALITIES

1. Introduction. In this chapter we establish the irrationality of some well-known numbers of elementary mathematics. The general criteria for irrationality given in Theorems 1.7 and 1.9 will not suffice for these numbers, and so in the next section we introduce a technique of Diophantine approximation. However, before getting into that, we set forth one more elementary criterion for irrationality, the Gaussian generalization of the usual proof that $\sqrt{2}$ is irrational.

THEOREM 2.1. *If the real number x satisfies an equation*

$$x^n + c_1 x^{n-1} + \cdots + c_n = 0$$

with integral coefficients, then x is either an integer or an irrational number.

Proof. Suppose that the number x is rational, say $x = a/b$, where the integers a and $b > 0$ satisfy $(a, b) = 1$. Then we would have

$$a^n = -b(c_1 a^{n-1} + c_2 a^{n-2} b + \cdots + c_n b^{n-1}).$$

If $b > 1$, then any prime divisor p of b would divide a^n, and so p would divide a by the fundamental theorem of arithmetic (cf. glossary). But this contradicts the condi-

15

tion $(a, b) = 1$, and so $b = 1$, which establishes that x, if rational, is an integer.

The irrationality of $\sqrt{2}$ is the particular case of Theorem 2.1 where the equation considered is $x^2 - 2 = 0$. More generally, by applying Theorem 2.1 to the equation $x^n - m = 0$, we obtain the following result.

COROLLARY 2.2. *If m is a positive integer which is not the n-th power of an integer, then $\sqrt[n]{m}$ is irrational.*

2. The trigonometric functions and π. Next we establish the irrationality of the trigonometric functions for non-zero rational values of the arguments. It will be convenient to have two lemmas.

LEMMA 2.3. *If $h(x) = x^n g(x)/n!$ where $g(x)$ is a polynomial with integral coefficients, then $h^{(j)}(0)$, the j-th derivative of $h(x)$ evaluated at $x = 0$, is an integer for $j = 0, 1, 2, \cdots$. Moreover, with the possible exception of the case $j = n$, the integer $h^{(j)}(0)$ is divisible by $n + 1$; no exception need be made in the case $j = n$ if $g(x)$ has the factor x: i.e., if $g(0) = 0$.*

Proof. We may write

$$h^{(j)}(0) = c_j(j!)/n!$$

where c_j is the coefficient of x^j in the polynomial $x^n g(x)$, so that c_j is an integer by hypothesis. For $j < n$ we have $c_j = 0$. For $j > n$ it is apparent that $h^{(j)}(0)$ is divisible by $n + 1$. In case $j = n$ we have $h^{(n)}(0) = c_n$, and, if $g(0) = 0$, then $c_n = 0$.

LEMMA 2.4. *If $f(x)$ is a polynomial in $(r - x)^2$, then $f^{(j)}(r) = 0$ for any odd integer j.*

Proof. For odd j, $f^{(j)}(x)$ is seen to be a polynomial in odd powers only of $(r - x)$, and the result follows.

THEOREM 2.5.   *For any rational number $r \neq 0$, $\cos r$ is irrational.*

Proof.   Since $\cos(-r) = \cos r$, it will suffice to prove this for positive $r = a/b$, where a and b are positive integers.   Define

$$(2.1) \quad f(x) = \frac{x^{p-1}(a - bx)^{2p}(2a - bx)^{p-1}}{(p-1)!}$$

$$= \frac{(r - x)^{2p}\{r^2 - (r - x)^2\}^{p-1}b^{3p-1}}{(p-1)!},$$

where p is an odd prime to be specified.   Note that, for $0 < x < r$,

$$(2.2) \quad 0 < f(x) < \frac{r^{2p}\{r^2\}^{p-1}b^{3p-1}}{(p-1)!} = \frac{r^{4p-2}b^{3p-1}}{(p-1)!}.$$

Next, using all the even derivatives of $f(x)$, we define

$$(2.3) \quad F(x) = f(x) - f^{(2)}(x) + f^{(4)}(x) - f^{(6)}(x)$$
$$+ \cdots - f^{(4p-2)}(x).$$

Thus we have

$$\frac{d}{dx}\{F'(x)\sin x - F(x)\cos x\}$$
$$= F^{(2)}(x)\sin x + F(x)\sin x = f(x)\sin x,$$

and

$$(2.4) \quad \int_0^r f(x)\sin x\, dx = F'(r)\sin r - F(r)\cos r + F(0).$$

Now $f(x)$ is a polynomial in $(r - x)^2$, and so by (2.3) and Lemma 2.4 it follows that $F'(r) = 0$.   But also $f(x)$ has the form of $h(x)$ in Lemma 2.3, with n here replaced by $p - 1$.   Thus we observe that $f^{(j)}(0)$ is an integer for every value of j, and moreover that $f^{(j)}(0)$ is a multiple of p unless $j = p - 1$. By direct calculation from (2.1) we have

$$f^{(p-1)}(0) = a^{2p}(2a)^{p-1}.$$

The odd prime p will be chosen to satisfy $p > a$, and, since a is a positive integer, it follows that $f^{(p-1)}(0)$ is not divisible by p. Thus we conclude that $F(0)$ is an integer not divisible by p; say $F(0) = q$ with $(p, q) = 1$.

Continuing our examination of the right side of (2.4), we now study $F(r)$. The definition (2.1) implies that

$$f(r - x) = \frac{x^{2p}\{r^2 - x^2\}^{p-1}b^{3p-1}}{(p - 1)!}$$

$$= \frac{x^{2p}\{a^2 - b^2x^2\}^{p-1}b^{p+1}}{(p - 1)!}.$$

Thus $f(r - x)$ has the form of $h(x)$ in Lemma 2.3 with n replaced by $p - 1$ and $g(x)$ replaced by

$$x^{p+1}\{a^2 - b^2x^2\}^{p-1}b^{p+1}.$$

Hence by Lemma 2.3 we see that $f^{(j)}(r)$ is, for every j, an integer divisible by p, and so $F(r)$ has the form pm where m is some integer.

We assume that $\cos r$ is rational; say $\cos r = d/k$, with integers d and $k > 0$. In view of this and our analysis of $F'(r)$, $F(0)$, and $F(r)$, we can rewrite (2.4) as

$$(2.5) \qquad k\int_0^r f(x) \sin x \, dx = -pmd + kq.$$

From this we shall get a contradiction by choosing the prime p sufficiently large so that the right side of (2.5) is a non-zero integer, and so that the left side is a value between $+1$ and -1. We have already specified that p should satisfy $p > a$, and now we add the requirement

$p > k$. Thus p is not a divisor of kq, and it follows that $-pmd + kq$ is a non-zero integer.

Turning to the left member of (2.5), we use (2.2) to write

$$|k\int_0^r f(x) \sin x \, dx| < kr\frac{r^{4p-2}b^{3p-1}}{(p-1)!} = kr^3b^2\frac{\{r^4b^3\}^{p-1}}{(p-1)!}$$

$$= c_1 c_2^{p-1}/(p-1)!$$

where the constants $c_1 = kr^3b^2$ and $c_2 = r^4b^3$ are independent of p. Now, as p tends to infinity, $c_1 c_2^{p-1}/(p-1)!$ tends to zero, and so we can choose p sufficiently large so that the left member of (2.5) lies between $+1$ and -1. This completes the proof of Theorem 2.5.

The above method of proof will be used several times in this chapter. Before proceeding, let us note the nature of the method. The heart of the proof lies in equation (2.4). Since the left member of (2.4) is arbitrarily small, and since $F'(r) = 0$, what equation (2.4) gives us is a very good rational approximation to $\cos r$: namely $F(0)/F(r)$. Now, if $\cos r$ were rational with denominator k, so would $-F(r)\cos r + F(0)$ be rational with denominator k, and not zero because of the divisibility properties of the arbitrarily large prime p. Thus $-F(r)\cos r + F(0)$ would have absolute value at least $1/k$, which is impossible since the left member of (2.4) can be made arbitrarily close to zero.

COROLLARY 2.6. π *is irrational.*

Proof. If π were rational, then $\cos \pi$ would be irrational by Theorem 2.5, whereas $\cos \pi = -1$.

Alternative proof. Instead of inferring the irrationality of π from Theorem 2.5, we can use a straightforward argument which avoids some of the complications of the theorem. Moreover the proof which we are about to give establishes the slightly stronger result that π^2 is irrational. Define

(2.6) $$f(x) = \frac{x^n(1-x)^n}{n!},$$

where n is a positive integer to be specified. Note that, for $0 < x < 1$,

(2.7) $$0 < f(x) < \frac{1}{n!}.$$

By Lemma 2.3, $f^{(j)}(0)$ is an integer for every value of j. Since $f(1-x) = f(x)$, it follows that $f^{(j)}(1)$ is also an integer for every j. Assume that $\pi^2 = a/b$, where a and b are positive integers, and define

$$F(x) = b^n\{\pi^{2n}f(x) - \pi^{2n-2}f^{(2)}(x) + \pi^{2n-4}f^{(4)}(x)$$
$$- \cdots + (-1)^n f^{(2n)}(x)\},$$

so that $F(0)$ and $F(1)$ are integers. We see that

$$\frac{d}{dx}\{F'(x)\sin \pi x - \pi F(x)\cos \pi x\}$$
$$= \{F^{(2)}(x) + \pi^2 F(x)\}\sin \pi x$$
$$= b^n\pi^{2n+2}f(x)\sin \pi x$$
$$= \pi^2 a^n f(x)\sin \pi x,$$

and so

(2.8) $$\pi a^n \int_0^1 f(x)\sin \pi x\, dx$$

$$= \left[\frac{F'(x)\sin \pi x}{\pi} - F(x)\cos \pi x\right]_0^1 = F(1) + F(0).$$

Now $F(1) + F(0)$ is an integer, but from (2.7) we have

$$0 < \pi a^n \int_0^1 f(x)\sin \pi x\, dx < \frac{\pi a^n}{n!} < 1$$

for sufficiently large n. Thus we have a contradiction in a simpler way than was the case with equation (2.5): it is simpler because we know that the integral in (2.8) is positive.

COROLLARY 2.7.   *The trigonometric functions are irrational at non-zero rational values of the arguments.*

Proof.   If sin r were rational for rational $r \neq 0$, so also would $1 - 2 \sin^2 r = \cos 2r$ be rational, contrary to Theorem 2.5. Similarly if tan r were rational, so would

$$\cos 2r = \frac{1 - \tan^2 r}{1 + \tan^2 r}$$

be rational, again in contradiction to the theorem. Also csc r, sec r and ctn r are irrational since they are the reciprocals of irrational numbers.

This proof of Corollary 2.7 establishes a little more: namely that $\sin^2 r$, $\cos^2 r$, etc., are irrational for $r \neq 0$.

We have derived the irrationality of sin r and tan r from that of cos r. It is interesting that cos r seems to be necessarily the basic function in this process. That is, whereas the irrationality of sin r and tan r is implied at once by that of cos r, it appears to be not possible in such a simple way to infer the irrationality of cos r from that of either sin r or tan r. This is the reason for our starting with the cosine function in Theorem 2.5.

COROLLARY 2.8.   *Any non-zero value of an inverse trigonometric function is irrational for a rational value of the argument.*

Proof.   Let r be rational, and consider one of the values of arc cos r, say arc cos $r = \rho$, and assume that ρ is a non-zero rational number. Then cos $\rho = r$, contrary to Theorem 2.5. A similar proof holds for the other functions.

3. The hyperbolic, exponential, and logarithmic functions.

THEOREM 2.9. *The hyperbolic functions are irrational for non-zero rational values of the arguments.*

Proof. Paralleling the proof of Theorem 2.5, we begin with cosh r where $r = a/b > 0$. We define $f(x)$ as in (2.1); however, in the present proof we do not require that p be a prime but merely a positive integer. Equations (2.3) and (2.4) are replaced by

$$F(x) = f(x) + f^{(2)}(x) + f^{(4)}(x) + f^{(6)}(x) + \cdots + f^{(4p-2)}(x)$$

and

$$(2.9) \quad \int_0^r f(x) \sinh x \, dx$$

$$= [F(x) \cosh x - F'(x) \sinh x]_0^r$$

$$= F(r) \cosh r - F'(r) \sinh r - F(0).$$

This is analogous to equation (2.4), but in the present case the integral in (2.9) is positive because of (2.2) and the fact that sinh x is positive for $0 < x < r$. (Knowing that the integral is positive enables us to give a shorter proof than in Theorem 2.5.) We apply Lemma 2.3 to $f(x)$ and $f(r - x)$ to conclude that $f^{(j)}(0)$ and $f^{(j)}(r)$ are integers for every j. Thus $F(0)$ and $F(r)$ are integers. Also $F'(r) = 0$ by Lemma 2.4. If we assume that cosh $r = d/k$, with $k > 0$, then (2.9) can be written as

$$k \int_0^r f(x) \sinh x \, dx = d \, F(r) - k \, F(0).$$

The right side is an integer, but by (2.2) we see that

$$0 < k \int_0^r f(x) \sinh x \, dx < kr \frac{r^{4p-2}b^{3p-1}}{(p-1)!} \cdot \frac{e^r - e^{-r}}{2}$$

$$= \frac{kr^3b^2(e^r - e^{-r})}{2} \cdot \frac{\{r^4b^3\}^{p-1}}{(p-1)!} < 1$$

for p sufficiently large. Thus we have a contradiction, so that $\cosh r$ is irrational.

Next we proceed from $\cosh r$ to $\sinh r$ and $\tanh r$ along the lines of the proof of Corollary 2.7, using now the identities

$$\cosh 2r = 2 \sinh^2 r + 1 = \frac{1 + \tanh^2 r}{1 - \tanh^2 r}.$$

An analogous result to Corollary 2.8 holds for the inverse hyperbolic functions; we omit the details.

THEOREM 2.10. *If r is rational and $r \neq 0$, then e^r is irrational. If r is rational and positive, and $r \neq 1$, then $\log r$ is irrational.*

The two statements are equivalent since $r = e^{\log r} = \log e^r$. Or, to say it another way, the two statements are related in the same way as are Corollaries 2.7 and 2.8, with inverse functions involved. It suffices therefore to establish the irrationality of e^r, and we give two proofs.

First proof. If e^r were rational, so would be its reciprocal e^{-r}, and so also $(e^r + e^{-r})/2$. But this is $\cosh r$, and so we have a contradiction to Theorem 2.9.

Second proof. This is a direct approach not involving hyperbolic functions. If e^r were rational, so would its integral powers be rational; so it suffices to consider only the cases where r is an integer. And, if $r < 0$, we may deal with e^{-r}; so we may presume that r is a positive integer. Assume that $e^r = a/b$, with $b > 0$.

We define $f(x)$ as in (2.6), and $F(x)$ by the equation

$$F(x) = r^{2n} f(x) - r^{2n-1} f'(x) + r^{2n-2} f^{(2)}(x)$$
$$- \cdots - r f^{(2n-1)}(x) + f^{(2n)}(x),$$

so that $F(0)$ and $F(1)$ are integers. We see that

$$\frac{d}{dx} \{e^{rx} F(x)\} = e^{rx}\{r F(x) + F'(x)\} = r^{2n+1} e^{rx} f(x),$$

and

$$br^{2n+1} \int_0^1 e^{rx} f(x)\, dx = b[e^{rx} F(x)]_0^1 = a F(1) - b F(0).$$

The right member is an integer, but by (2.7) we have

$$0 < br^{2n+1} \int_0^1 e^{rx} f(x)\, dx < \frac{br^{2n+1} e^r}{n!} = bre^r \frac{(r^2)^n}{n!} < 1$$

for n sufficiently large. Thus we have the usual contradiction.

The theorem just proved is an assertion about natural logarithms, and we now prove a similar result for common logarithms. The following simple result is stated for logarithms to base 10, but admits of generalization to any rational base.

THEOREM 2.11. *For any positive rational r, $\log_{10} r$ is irrational unless $r = 10^n$ for some integer n.*

Proof. If $\log_{10} r$ is rational, so is $\log_{10} r^{-1}$, and hence we may take $r > 1$. Let us write $r = a/b$ with positive integers a and b satisfying $(a, b) = 1$. Assume that $\log_{10} r = c/d$ with positive integers c and d satisfying $(c, d) = 1$. Then $10^c b^d = a^d$, which implies that $b = 1$ and $10^c = a^d$. Thus a must have the form $2^u 5^v$ with positive integers u and v, and consequently $c = ud$. But $(c, d) = 1$, so that $d = 1$ and $r = a/b = a = a^d = 10^c$.

If this theorem is generalized from logarithmic base 10 to any rational base $b \neq 1$, it becomes almost tautological; thus: if r and $b \neq 1$ are positive rationals, then $\log_b r$ is irrational unless there exist integers m and n such that $r^m = b^n$.

We conclude with a result which, although somewhat out of keeping with the title of this chapter, is included at this point because the necessary techniques of proof have been fully developed. A much more general result, Theorem 9.1, is given later.

THEOREM 2.12.   *e satisfies no relation of the form*

$$(2.10) \qquad a_m e^m + a_{m-1} e^{m-1} + \cdots + a_1 e + a_0 = 0$$

having integral coefficients not all zero. (Stated otherwise, e is a transcendental number.)

Proof. There is no loss of generality in presuming that $a_0 \neq 0$. We define

$$f(x) = \frac{x^{p-1}(x-1)^p (x-2)^p (x-3)^p \cdots (x-m)^p}{(p-1)!}$$

and

$$F(x) = f(x) + f'(x) + f^{(2)}(x) + f^{(3)}(x) \\ + \cdots + f^{(mp+p-1)}(x),$$

where p is an odd prime to be specified. For $0 < x < m$ we have

$$(2.11) \qquad |f(x)| < \frac{m^{p-1} m^p m^p \cdots m^p}{(p-1)!} = \frac{m^{mp+p-1}}{(p-1)!}.$$

We readily verify that

$$\frac{d}{dx} \{e^{-x} F(x)\} = e^{-x} \{F'(x) - F(x)\} = -e^{-x} f(x)$$

and

$$a_j \int_0^j e^{-x} f(x)\, dx = a_j [-e^{-x} F(x)]_0^j = a_j F(0) - a_j e^{-j} F(j).$$

We multiply by e^j, then sum over the values $j = 0, 1, \cdots,$ m, and get, by use of (2.10),

$$(2.12) \quad \sum_{j=0}^{m} a_j e^j \int_0^j e^{-x} f(x) \, dx = -\sum_{j=0}^{m} a_j F(j)$$

$$= -\sum_{j=0}^{m} \sum_{i=0}^{mp+p-1} a_j f^{(i)}(j).$$

Now the application of Lemma 2.3 to $f(x)$, $f(x + 1)$, $f(x + 2)$, $\cdots$, $f(x + m)$, with n replaced by $p - 1$ in each case, shows that $f^{(i)}(j)$ is an integer for all values of i and j in the above sum. Even more, it shows that $f^{(i)}(j)$ is an integer divisible by p except for the single case where $j = 0$ and $i = p - 1$. A direct calculation from the definition of $f(x)$ establishes that

$$f^{(p-1)}(0) = (-1)^p (-2)^p \cdots (-m)^p.$$

Thus $f^{(p-1)}(0)$ is not divisible by p if we choose $p > m$. Furthermore, if we choose $p > |a_0|$, we see that the right member of (2.12) consists of a sum of multiples of p with one exception, namely $-a_0 f^{(p-1)}(0)$. Thus the sum on the right of (2.12) is a non-zero integer. But the left member of (2.12) satisfies the inequality, by (2.11),

$$\left| \sum_{j=0}^{m} a_j e^j \int_0^j e^{-x} f(x) \, dx \right| \leq \sum \left| a_j e^j \int_0^j e^{-x} f(x) \, dx \right|$$

$$\leq \sum |a_j| e^j j \cdot 1 \cdot \frac{m^{mp+p-1}}{(p-1)!}$$

$$\leq \left\{ \sum |a_j| \right\} e^m m \frac{m^{mp+p-1}}{(p-1)!}$$

$$\leq \left\{ \sum |a_j| \right\} e^m \frac{(m^{m+2})^{p-1}}{(p-1)!} < 1,$$

provided p is chosen sufficiently large. Thus we have a contradiction, and the theorem is proved.

Notes on Chapter 2

The irrationality of π (Corollary 2.6) was first proved by J. H. Lambert in 1761 by means of continued fractions. An exposition of this approach to the topic is given in G. Chrystal, *Algebra*, vol. II, Chapter 34.

The problem of "squaring the circle" cannot be settled by the irrationality of π; it was solved by F. Lindemann in 1882 by the proof of the transcendence of π. This is discussed in § 4 and § 5 of Chapter 9.

The basic technique of this chapter, and also of Chapter 9, is due to C. Hermite, *Compt. Rend. Acad. Sci. Paris*, **77** (1873), 18–24, 74–79, 285–293. There have been various refinements and extensions of the technique by later writers. Complete references up to 1934 are to be found in Chapter IV of J. F. Koksma, *Diophantische Approximationen*.

The particular form of Hermite's method used in Theorems 2.5 and 2.9 developed from a proof of the irrationality of π given by us in *Bull. Amer. Math. Soc.*, **53** (1947), 509. This was extended to π^2 by Y. Iwamoto, *J. Osaka Inst. Sci. Tech.* (1) **1** (1949), 147–148; to the exponential function by Z. Butlewski, *Colloq. Math.*, **1** (1948), 197–198, and by J. F. Koksma, *Nieuw Arch. Wiskunde* (2) **23** (1949), 39; and to solutions of certain differential equations by R. Rado, *J. London Math. Soc.*, **23** (1948), 267–271. Other simple approaches to π and $\tan r$ have been given by J. Popken, *Norsk. Mat. Tidsskr.*, **26** (1944), 66–70; *Math. Centrum Amsterdam*, rapport **ZW1948-014**, 5 pp. (1948); and R. Breusch, *Amer. Math. Monthly*, **61** (1954), 631–632.

The proof of Theorem 2.5, which we believe is new, was improved after discussion of an earlier version with Professor Saunders Mac-Lane. The unified treatment of π and e^r in Corollary 2.6 and Theorem 2.10 by use of (2.6) follows Hardy and Wright, *The Theory of Numbers*, Chapter IV. The proof of Theorem 2.12 is based on that of A. Hurwitz, *Math. Annalen*, **43** (1893), 220–222, or *Math. Werke*, *Basel*, **II** (1933), 134–135.

CHAPTER 3

CERTAIN ALGEBRAIC NUMBERS

1. Introduction. The trigonometric functions for rational values of the argument or angle were discussed in the preceding chapter. Now we examine the values of the trigonometric functions for those angles that are rational multiples of π: i.e., for angles that are rational when measured in degrees. Apart from trivial exceptions these values are irrational numbers, but they are also algebraic numbers whose degrees we shall calculate. We now state the prerequisite material needed up to § 4 of this chapter; one or two additional results are required in § 5 and are given there.

An *algebraic number* is one that satisfies some equation of the form

$$(3.1) \qquad x^n + a_1 x^{n-1} + \cdots + a_n = 0,$$

with rational coefficients. A polynomial having leading coefficient 1, such as that in (3.1), is called *monic*. Any algebraic number α satisfies a unique monic polynomial equation of least degree, called the *minimal polynomial* of α and the *degree* of α. Thus rational numbers coincide with algebraic numbers of degree 1. Also the minimal polynomial of α is irreducible over the rational numbers, and it is a divisor of any other polynomial with rational coefficients having α as a zero, and it is the only monic polynomial having these properties. Proofs of these simple properties of algebraic numbers can be found in the Carus Monograph of Harry Pollard, *The Theory of Algebraic Numbers*, pp. 35–36.

If an algebraic number α satisfies some equation (3.1) with rational integral coefficients, we say that α is an *algebraic integer*. For example, of the algebraic numbers $\sqrt{3}$ and $\sqrt{3}/2$, only the first is an algebraic integer. The minimal polynomial of an algebraic integer is also monic with integral coefficients. This can be proved by applying *Gauss's lemma*, a form of which we now state: if a polynomial $f(x)$ with rational integral coefficients can be factored into two polynomials with rational coefficients, say $f(x) = g(x)\, h(x)$, then there is a non-zero rational number r such that $r\, g(x)$ and $r^{-1}\, h(x)$ have rational integral coefficients; for proof see Pollard, Theorem 3.7. We shall use later a special case of Gauss's lemma, namely: if $f(x) = g(x)\, h(x)$, and if $f(x)$ and $g(x)$ are monic with integral coefficients, then also $h(x)$ is monic with integral coefficients.

A complex number that is not algebraic is called *transcendental*. We shall establish the transcendental character, or transcendence, of some classes of numbers in Chapters 7, 9, and 10; the number e was treated in Theorem 2.12.

Our purpose in this chapter is to prove the algebraic nature of the trigonometric functions at arguments that are rational multiples of π, and to determine the degrees of these algebraic numbers. This necessitates looking into the simplest parts of the theory of cyclotomic polynomials, a topic of interest in its own right.

It may be noted very simply that the trigonometric functions are algebraic numbers at arguments that are rational multiples of π. First consider $\cos\theta$ with $\theta =$

$2\pi k/n$, where k and n are relatively prime integers. We have $(\cos\theta + i\sin\theta)^n = 1$ by De Moivre's theorem. Writing the binomial expansion of the left side of this equation, we equate the real parts to obtain a polynomial equation in $\cos\theta$ and $\sin\theta$, the latter occurring to even powers only. Replacing $\sin^2\theta$ by $1 - \cos^2\theta$, we get an algebraic equation satisfied by $\cos\theta$. (This equation does not vanish identically: e.g., the coefficient of $\cos^n\theta$ is 2^{n-1}.) Also $\sin\theta$ is algebraic, since $\sin\theta = \cos(\theta - \pi/2)$. Furthermore, if the imaginary parts are equated in the expansion of $(\cos\theta + i\sin\theta)^n = 1$, and if the result is divided through by $\cos^n\theta$, the equation obtained establishes that $\tan\theta$ is algebraic.

2. Further background material. The factorization of $x^n - 1$ in the field (cf. glossary) of complex numbers,

$$x^n - 1 = \prod_{k=0}^{n-1} (x - e^{2\pi i k/n})$$

$$= \prod_{k=0}^{n-1} \left(x - \cos\frac{2\pi k}{n} - i\sin\frac{2\pi k}{n} \right),$$

is well known, the zeros of the polynomial being the nth roots of unity. In §3 we shall determine for later use the complete factorization of $x^n - 1$ in the field of rational numbers for any positive integer n. Among the nth roots of unity a special role is played by the so-called *primitive roots*, namely $e^{2\pi i k/n}$ for values of k such that $(k, n) = 1$. There are $\phi(n)$ primitive nth roots of unity, $\phi(n)$ being the *Euler* function with value

$$\phi(n) = n\left(1 - \frac{1}{p_1}\right)\left(1 - \frac{1}{p_2}\right) \cdots \left(1 - \frac{1}{p_r}\right),$$

where $p_1, p_2, \cdots, p_r$ are the distinct prime factors of n. If ω is any primitive nth root of unity, so also is ω^j for any exponent j satisfying $(j, n) = 1$.

For any prime p, the numbers 0, 1, 2, $\cdots$, $p - 1$ constitute a finite field: i.e., a field with a finite number of elements, addition and multiplication being defined modulo p. (This assertion can be readily verified, in case the reader is unfamiliar with these concepts, from the definition of "field" given in the glossary. The only point of any difficulty whatsoever is the multiplicative inverse of a non-zero element a: the congruence $ax \equiv 1 \pmod{p}$ is known to have a unique solution x among the values 1, 2, $\cdots$, $p - 1$, and this solution is the inverse of a, denoted by a^{-1}.) This finite field is denoted by J_p; and by $J_p[x]$ we shall mean the set of all polynomials in x with coefficients in J_p. The degree of a non-zero polynomial is the exponent of the highest power of x, so that the non-constant polynomials are those of degree at least one.

We say that the polynomial $g(x)$ in $J_p[x]$ is a divisor or factor of $f(x)$ in case there is a polynomial $q(x)$ such that $f(x) = g(x)\, q(x)$. Any non-constant polynomial is said to be *irreducible* if it cannot be factored into two non-constant polynomials in $J_p[x]$. Either any non-constant polynomial is irreducible or it can be factored into irreducible polynomials in $J_p[x]$. One basic result which we shall need is that this factorization into irreducible polynomials is unique. The proof of this is so similar to the proof of the unique factorization theorem for polynomials over the rational numbers that we merely sketch the broad steps: the details are quite analogous to those given for example by Pollard, pp. 24–26.

First, if $f(x)$ and $g(x)$ are polynomials in $J_p[x]$ of degrees n and m, respectively, and if $n \geqq m$, then there is an element c in J_p such that the expression

$$f(x) - cx^{n-m}\, g(x)$$

is zero or a polynomial of degree less than n. From this we can get a Euclidean algorithm for $J_p[x]$, namely:

Given any two polynomials $f(x)$ and $g(x) \neq 0$, there exist polynomials $q(x)$ and $r(x)$ such that

$$f(x) = g(x)\, q(x) + r(x)$$

where $r(x) = 0$ or $r(x)$ is of lower degree than $g(x)$. Next, this algorithm can be used to prove that, given any two non-zero polynomials $f(x)$ and $g(x)$ having no common factor of degree one or more, then there exist polynomials $s(x)$ and $t(x)$ in $J_p[x]$ such that

$$f(x)\, s(x) + q(x)\, t(x) = 1.$$

Now let $m(x)$ be an irreducible polynomial which is a divisor of a product of two polynomials $u(x)\, v(x)$ but is not a divisor of $u(x)$ alone: then $m(x)$ is a divisor of $v(x)$. This chain of results leads to the unique factorization property, which we state as a lemma for reference.

LEMMA 3.1. *Any non-constant polynomial $f(x)$ in $J_p[x]$ can be factored into a product of monic irreducible polynomials*

$$f(x) = cm_1(x)\, m_2(x) \cdots m_r(x)$$

in one and only one way apart from the order of the factors, where c is the leading coefficient of $f(x)$.

LEMMA 3.2. *If $g(x)$ is any polynomial in $J_p[x]$, then $\{g(x)\}^p = g(x^p)$.*

Proof. If $g(x)$ is a constant polynomial c, this result is Fermat's theorem to the effect that $c^p \equiv c \pmod{p}$. We can prove the general result by using induction on the degree of the polynomial $g(x)$. Suppose that the lemma holds for any polynomial of degree less than n. Then, if $g(x)$ has degree n, it can be written in the form

$$g(x) = ax^n + g_1(x),$$

where $g_1(x)$ has degree less than n. Then we have

$$\{g(x)\}^p = (ax^n)^p + \binom{p}{1}(ax^n)^{p-1} g_1(x)$$

$$+ \binom{p}{2}(ax^n)^{p-2}\{g_1(x)\}^2 + \cdots + \{g_1(x)\}^p.$$

But by the induction hypothesis $\{g_1(x)\}^p = g_1(x^p)$, and it is well known that the binomial coefficient $\binom{p}{j}$ is a multiple of p for any integer j in the range $0 < j < p$. Hence the above equation implies that

$$\{g(x)\}^p = a^p x^{np} + g_1(x^p) = ax^{np} + g_1(x^p) = g(x^p).$$

Finally we state a lemma concerning the following formal type of differentiation of polynomials of $J_p[x]$. The derivative of

$$f(x) = a_0 x^n + a_1 x^{n-1} + \cdots a_n$$

is defined as

$$f'(x) = na_0 x^{n-1} + (n-1)a_1 x^{n-2} + \cdots a_{n-1}.$$

LEMMA 3.3.   *If $f(x)$ and $g(x)$ are any polynomials in $J_p[x]$, then*

$$\{f(x)\,g(x)\}' = f(x)\,g'(x) + f'(x)\,g(x).$$

A proof of this lemma can be readily constructed, for example, by expanding both sides of the equation; so we omit the details.

3. The factorization of $x^n - 1$.   By $F_n(x)$ we shall mean the *cyclotomic* polynomial of degree $\phi(n)$ whose zeros are the primitive nth roots of unity; thus

$$(3.2) \qquad F_n(x) = \prod_{\substack{k=0 \\ (k,\,n)=1}}^{n-1} (x - e^{2\pi i k/n}).$$

The gist of this section is that the equation in the next theorem gives the complete factorization of $x^n - 1$ into polynomials with rational coefficients, or for that matter with rational integral coefficients.

THEOREM 3.4. *For $n \geqq 1$,*

$$x^n - 1 = \prod_{d \mid n} F_d(x),$$

the product being over all divisors d of n.

Proof. Any zero of $x^n - 1$, say $e^{2\pi i k/n}$, is a primitive dth root of unity for $d = n/(k, n)$. Hence any linear factor of $x^n - 1$ occurs in some $F_d(x)$. Conversely any linear factor of $F_d(x)$ is of the form $x - \omega$ where ω is a primitive dth root of unity. But any dth root of unity is also an nth root of unity, so that any linear factor of any $F_d(x)$ is a factor of $x^n - 1$. The proof is completed by noting that just as the linear factors of $x^n - 1$ are distinct, so are the linear factors of $\Pi F_d(x)$, since a primitive dth root of unity is not also a primitive d_1th root of unity, d and d_1 being assumed unequal.

THEOREM 3.5. *For $n \geqq 1$, $F_n(x)$ is a monic polynomial of degree $\phi(n)$ with rational integral coefficients.*

Proof. The polynomial $F_n(x)$ is monic by the equation (3.2), and its degree is $\phi(n)$ since there are exactly that many primitive nth roots of unity. We use induction on n to prove that the coefficients are integers. For $n = 1$ we have $F_1(x) = x - 1$. Assuming that all $F_m(x)$ for $m < n$ have integral coefficients, define

$$g_1(x) = \frac{x^n - 1}{F_n(x)} = \prod_{\substack{d \mid n \\ d < n}} F_d(x).$$

Then $g_1(x)$, a product of monic polynomials with integral coefficients, is monic with integral coefficients. The application of Gauss's lemma to the equation $x^n - 1 = g_1(x) F_n(x)$ establishes that $F_n(x)$ has integral coefficients.

Next, let ω be a primitive nth root of unity, and define $f(x)$ as the minimal polynomial of ω. As we shall see, it turns out that $f(x)$ so defined and $F_n(x)$ defined in (3.2) are identical. Now $f(x)$ is monic by definition, and has integral coefficients since ω, being a root of $x^n - 1 = 0$, is an algebraic integer.

LEMMA 3.6. *If p is any prime such that $(p, n) = 1$, and if ρ is any root of $f(x) = 0$, then ρ^p is also a root.*

Proof. Since ω is by definition a zero of both $f(x)$ and $F_n(x)$, it follows that the minimal polynomial $f(x)$ is a divisor of $F_n(x)$, so that all the roots of $f(x) = 0$ are primitive nth roots of unity. Furthermore, since $f(x)$ is irreducible, it is the minimal polynomial not only of ω but also of each of its zeros. By the fundamental theorem on symmetric polynomials (cf. glossary) the pth powers of the roots of $f(x) = 0$ are roots of a polynomial equation which is also monic with integral coefficients, and so ρ^p is also an algebraic integer.

Let $g(x)$, monic with rational integral coefficients, be the minimal polynomial of ρ^p. We establish that $g(x) = f(x)$, and from this the lemma follows at once. Suppose, on the contrary, that $f(x) \neq g(x)$. Then, being irreducible, $f(x)$ and $g(x)$ would have g.c.d. 1. But $g(x)$ is a divisor of $x^n - 1$ since they have the common zero ρ^p, an nth root of unity. Similarly $f(x)$ is a divisor of $x^n - 1$, so that the product $f(x) g(x)$ is a divisor of $x^n - 1$, say

$$(3.3) \qquad x^n - 1 = f(x) g(x) q(x).$$

Gauss's lemma implies that $q(x)$ is also monic with rational integral coefficients. Now ρ is a root of $g(x^p) = 0$, so that $f(x)$ is a divisor of $g(x^p)$, say $g(x^p) = f(x) Q(x)$, with $Q(x)$ again monic with integral coefficients.

We now interpret these equations in $J_p[x]$. By Lemma 3.2 the last equation can be written as $\{g(x)\}^p = f(x)\,Q(x)$. Let $k(x)$ be any irreducible polynomial factor of $f(x)$ in $J_p[x]$. Thus $k(x)$ is a divisor of $\{g(x)\}^p$, and by Lemma 3.1 it follows that $k(x)$ is a divisor of $g(x)$. Hence by equation (3.3), $x^n - 1$ is divisible by $\{k(x)\}^2$, say $x^n - 1 = s(x)\,\{k(x)\}^2$. Upon formal differentiation of this equation, we have, by Lemma 3.3,

$$nx^{n-1} = k(x)\,\{s'(x)\,k(x) + 2s(x)\,k'(x)\}.$$

Note that nx^{n-1} is not zero in $J_p[x]$ since $(n, p) = 1$. But the only irreducible polynomial factor of nx^{n-1} is x, so that $k(x) = x$, which is impossible since $k(x)$, being a factor of $f(x)$, is also a factor of $x^n - 1$ by (3.3). This contradiction establishes that $f(x) = g(x)$, and this completes the proof of Lemma 3.6.

THEOREM 3.7. *The cyclotomic polynomial $F_n(x)$ is irreducible over the field of rational numbers.*

Proof. We know that $f(x)$, as defined just before Lemma 3.6, is irreducible since it is a minimal polynomial. We shall establish that any primitive nth root of unity is a root of $f(x) = 0$, and it follows that $f(x) = F_n(x)$, which proves the theorem. Any primitive nth root of unity can be written as a power of the particular root ω which was used to define $f(x)$, say ω^t where $(t, n) = 1$. If the factorization of t into (not necessarily distinct) primes is $t = p_1 p_2 \cdots p_s$, then $(p_i, n) = 1$ for all subscripts i. Then, by Lemma 3.6, ω^{p_1} is a root of $f(x) = 0$, and by iteration of the argument $\omega^{p_1 p_2}$ is a root of $f(x) = 0$, and so by induction ω^t is a root of $f(x) = 0$.

4. Certain trigonometric values. The reciprocal of a primitive nth root of unity is also a primitive nth root of unity. Moreover, the reciprocals of a complete set of primitive nth roots of unity are the same numbers again in some order, and hence the polynomial $F_n(x)$ defined by (3.2) has the property that $x^{\phi(n)} F_n(x^{-1}) = F_n(x)$. Thus the coefficient of x^j in $F_n(x)$ equals the coefficient of $x^{\phi(n)-j}$, for any j in the range $0 \leqq j \leqq \phi(n)$. Let $n > 2$ so that $\phi(n)$ is even, say $\phi(n) = 2m$. Then $x^{-m} F_n(x)$ has the form

$$x^{-m} F_n(x) = (x^m + x^{-m}) + a_1(x^{m-1} + x^{1-m})$$
$$+ \cdots + a_{m-1}(x + x^{-1}) + a_m,$$

with integral coefficients.   Now the identity

$$x^k + x^{-k} = (x + x^{-1})(x^{k-1} + x^{1-k}) - (x^{k-2} + x^{2-k})$$

enables us to establish readily by mathematical induction that $x^k + x^{-k}$ is a monic polynomial in $x + x^{-1}$ of degree k with integral coefficients.   This analysis proves the first part of the following proposition.

LEMMA 3.8.   *Let* $n > 2$, *and define* $m = \phi(n)/2$. *Then* $x^{-m} F_n(x)$ *is a monic polynomial in* $x + x^{-1}$, *say* $\psi_n(x + x^{-1})$, *with rational integral coefficients. Also* $\psi_n(x)$ *is irreducible of degree* m.

Proof.   It is clear that $\psi_n(x)$ has degree m; so it remains to prove the irreducibility.   Suppose contrariwise that $\psi_n(x) = h_1(x) h_2(x)$, where h_1 and h_2 are non-constant polynomials in x, with h_1 say of degree r.   Then we could write

$$F_n(x) = x^m \psi_n(x + x^{-1})$$
$$= \{x^r h_1(x + x^{-1})\} \cdot \{x^{m-r} h_2(x + x^{-1})\},$$

and this contradicts Theorem 3.7.

THEOREM 3.9.   (D. H. Lehmer)   *If* $n > 2$ *and* $(k, n) = 1$, *then* $2 \cos 2\pi k/n$ *is an algebraic integer of degree* $\phi(n)/2$. *For positive* $n \neq 4$, $2 \sin 2\pi k/n$ *is an algebraic integer of degree* $\phi(n)$, $\phi(n)/4$, *or* $\phi(n)/2$ *according as* $(n,8) < 4$, $(n, 8) = 4$, *or* $(n, 8) > 4$.

Proof. Since $e^{2\pi ik/n}$ is a root of $F_n(x) = 0$, it follows that $2 \cos 2\pi k/n = e^{2\pi ik/n} + e^{-2\pi ik/n}$ is a root of $\psi_n(x) = 0$, and so is an algebraic integer of degree $\phi(n)/2$ by Lemma 3.8. Next we observe that

$$2 \sin 2\pi k/n = 2 \cos 2\pi(4k - n)/4n,$$

and so we investigate the lowest terms of the fraction $(4k - n)/4n$. This fraction (i) is in its lowest terms if n is odd, (ii) reduces to a fraction with denominator $2n$ if $n \equiv 2 \pmod 4$, and (iii) reduces to a fraction with denominator n or smaller if $n \equiv 0 \pmod 4$.

In case (i) we have, letting d denote the degree of the algebraic integer $2 \sin 2\pi k/n$, $d = \phi(4n)/2 = \phi(n)$. In case (ii) we have $d = \phi(2n)/2 = \phi(n)$. Case (iii) separates into two subcases as follows. If $n \equiv 0 \pmod 8$ then, k being odd, the fraction $(4k - n)/4n$ reduces to a fraction with denominator n, so that $d = \phi(n)/2$. On the other hand, if $n \equiv 4 \pmod 8$ the fraction $(4k - n)/4n$ reduces to one with denominator $n/4$ in case $k \equiv n/4 \pmod 4$ and denominator $n/2$ otherwise, and it is readily calculated that $d = \phi(n)/4$ in both instances.

The restriction $n \neq 4$ in the sine part of the theorem is needed to avoid having $(4k - n)/4n$ reduce to denominator 1 or 2, in order that the cosine result is always applicable.

5. Extension to the tangent. We now extend Lehmer's theorem to the case of $\tan 2\pi k/n$, an algebraic number whose degree we compute. However, the number $2 \tan 2\pi k/n$ is not always an algebraic integer, for example in the case $k = 1$, $n = 12$. Our procedure necessitates the use of a couple of results on algebraic number fields which, though fairly elementary, are more advanced than the ideas previously employed in this chapter. We state these results now. Let R denote the field of rational numbers, and $R(u)$ the extension of R by the algebraic number u. The degree of $R(u)$ over R is the same as the degree of u. The principal result required is that, if H is a finite extension of R, and K in turn a finite extension of H, then

$$[K:R] = [K:H][H:R],$$

where the notation $[K:R]$ denotes the degree of K over R. (For proof see Pollard, p. 51.) A special case of this result is that, if u and v are algebraic numbers of the same degree, and if $R(v)$ is a subfield of $R(u)$, then $R(u) = R(v)$. We write $d(u)$ for the degree of the algebraic number u.

LEMMA 3.10. $R(\cos 2\theta)$ *is a subfield of each of the fields* $R(\cos \theta)$, $R(\sin \theta)$ *and* $R(\tan \theta)$, *for any argument* θ *for which* $\tan \theta$ *exists.*

Proof.   This is clear from the identities $\cos 2\theta = 2 \cos^2 \theta - 1 = 1 - 2 \sin^2 \theta = 2/(1 + \tan^2 \theta) - 1$.

THEOREM 3.11.   *For* $n > 4$ *and* $(k, n) = 1$, *the degree of* $\tan 2\pi k/n$ *is* $\phi(n)$, $\phi(n)/2$, *or* $\phi(n)/4$, *according as* $(n, 8) < 4$, $(n, 8) = 4$, *or* $(n, 8) > 4$.

Proof.   *Case 1:* $(n, 8) < 4$. Write θ for $2\pi k/n$. Then Theorem 3.9 says that

$$d(\cos \theta) = d(\cos 2\theta) = \tfrac{1}{2}d(\sin \theta).$$

By Lemma 3.10 we conclude that

$$R(\cos \theta) = R(\cos 2\theta) \subset R(\sin \theta),$$

and hence $\tan \theta \in R(\sin \theta)$. This with Lemma 3.10 implies that

$$R(\cos \theta) = R(\cos 2\theta) \subset R(\tan \theta) \subset R(\sin \theta).$$

Also the degree $[R(\sin\,\theta):R(\cos\,\theta)] = 2$, so that either $R(\tan\,\theta) = R(\sin\,\theta)$ or $R(\tan\,\theta) = R(\cos\,\theta)$. But the latter is impossible since it would imply that $\sin\,\theta \in R(\cos\,\theta)$. Hence

$$R(\tan\,\theta) = R(\sin\,\theta) \quad \text{and} \quad d(\tan\,\theta) = d(\sin\,\theta) = \phi(n).$$

Case 2: $(n, 8) = 4$. The proof parallels that in case 1, with the roles of $\sin\,\theta$ and $\cos\,\theta$ interchanged. Specifically, we can write

$$d(\sin\,\theta) = d(\cos\,2\theta) = \tfrac{1}{2}d(\cos\,\theta),$$

$$R(\sin\,\theta) = R(\cos\,2\theta) \subset R(\cos\,\theta),$$

$$(3.4) \qquad\qquad \tan\,\theta \in R(\cos\,\theta).$$

We continue the argument with

$$R(\sin\,\theta) = R(\cos\,2\theta) \subset R(\tan\,\theta) \subset R(\cos\,\theta),$$

$$[R(\cos\,\theta):R(\sin\,\theta)] = 2,$$

and $R(\tan\,\theta) = R(\sin\,\theta)$ is impossible in this case since it leads to the conclusion $\cos\,\theta \in R(\sin\,\theta)$. Hence

$$R(\tan\,\theta) = R(\cos\,\theta) \quad \text{and} \quad d(\tan\,\theta) = d(\cos\,\theta) = \phi(n)/2.$$

Case 3: $(n, 8) = 8$. In this case we will establish that $R(\tan\,\theta) = R(\cos\,2\theta)$, from which it follows that

$$d(\tan\,\theta) = d(\cos\,2\theta) = \phi(n)/4.$$

Now $R(\cos\,2\theta) \subset R(\tan\,\theta)$ by Lemma 3.10, and so we prove that $R(\tan\,\theta) \subset R(\cos\,2\theta)$ or that $\tan\,\theta \in R(\cos\,2\theta)$, which we do by induction on j, 2^j being the highest power of 2 dividing n. First for $j = 3$, the smallest value of j to be considered since $(n, 8) = 8$, we note that 2θ is of the type treated in case 2, and formula (3.4) would now be written as $\tan\,2\theta \in R(\cos\,2\theta)$. It follows that $\sin\,2\theta \in R(\cos\,2\theta)$, and so

$$\tan \theta = \frac{\sin 2\theta}{1 + \cos 2\theta} \in R(\cos 2\theta).$$

For any $j > 3$ we assume by the induction hypothesis that $\tan 2\theta \in R(\cos 4\theta)$ and we must prove that $\tan \theta \in R(\cos 2\theta)$. Now $R(\cos 4\theta) \subset R(\cos 2\theta)$ by Lemma 3.10, and consequently

$$\tan 2\theta \in R(\cos 2\theta), \ \sin 2\theta \in R(\cos 2\theta),$$

$$\tan \theta = \frac{\sin 2\theta}{1 + \cos 2\theta} \in R(\cos 2\theta).$$

Thus the proof is complete, and we note the following consequence of Theorems 3.9 and 3.11.

COROLLARY 3.12. *If θ is rational in degrees, say $\theta = 2\pi r$ for some rational number r, then the only rational values of the trigonometric functions of θ are as follows:* $\sin \theta$, $\cos \theta$ $= 0$, $\pm\frac{1}{2}$, ± 1; $\sec \theta$, $\csc \theta = \pm 1$, ± 2; $\tan \theta$, $\cot \theta = 0$, ± 1.

Notes on Chapter 3

The central result, Theorem 3.9, was proved by D. H. Lehmer, *Amer. Math. Monthly*, **40** (1933), 165–166. The extension to the tangent function in Theorem 3.11 has not been given elsewhere, so far as we know. A proof of Corollary 3.12 independent of Theorems 3.9 and 3.11 was given by J. M. H. Olmsted, *Amer. Math. Monthly*, **52** (1945), 507–508. The topic is a recurring one in the popular literature: as examples we cite B. H. Arnold and Howard Eves, *Amer. Math. Monthly*, **56** (1949), 20–21; R. W. Hamming, *Amer. Math. Monthly*, **52** (1945), 336–337; E. Swift, *Amer. Math. Monthly*, **29** (1922), 404–405; R. S. Underwood, *Amer. Math. Monthly*, **28** (1921), 374–376.

CHAPTER 4

THE APPROXIMATION OF
IRRATIONALS BY RATIONALS

1. The problem. Given an irrational number α, it is clear that there are rational numbers h/k close to α, so that $|\alpha - h/k|$ is small. How small? Since by Theorem 1.5 the rationals are dense, we can choose h/k so that for any arbitrary positive ϵ we have $|\alpha - h/k| < \epsilon$. If we presume k positive, this can be written as $|k\alpha - h| < \epsilon k$. This inequality suggests, but does not solve, the problem of trying to select the integer k so that $k\alpha$ is arbitrarily close to an integer. Stated completely the problem is this: Given an irrational number α, can we find a positive integer k so that $k\alpha$ is arbitrarily close to the nearest integer; that is $|k\alpha - h| < \epsilon$? The answer is yes (Corollary 4.2), and we solve this and some related problems by a very simple method (the pigeon-hole method) in this chapter. Deeper results will be obtained in Chapter 6 by use of continued fractions.

THEOREM 4.1. *For any irrational α there exist infinitely many rationals h/k such that $|\alpha - h/k| < 1/k^2$.*

Proof. Let n be any positive integer. Consider the $n + 1$ real numbers

$$(4.1) \qquad 0, \ \alpha - [\alpha], \ 2\alpha - [2\alpha], \ \cdots, \ n\alpha - [n\alpha],$$

and their distribution in the n intervals

$$j/n \leqq x < (j + 1)/n, \qquad j = 0, 1, 2, \cdots, n - 1$$

These n intervals cover the unit interval $0 \leq x < 1$ and hence contain the $n + 1$ values (4.1). Hence two of the numbers (4.1) lie in the same interval; this is the *pigeon-hole principle*. Call these numbers $n_1\alpha - [n_1\alpha]$ and $n_2\alpha - [n_2\alpha]$ with $0 \leq n_1 < n_2 \leq n$. Since the intervals are of length $1/n$ and are not closed at both ends, the difference between the two numbers is less than $1/n$; thus

$$|n_2\alpha - [n_2\alpha] - n_1\alpha + [n_1\alpha]| < 1/n.$$

Write k for the positive integer $n_2 - n_1$, and h for $[n_2\alpha] - [n_1\alpha]$; so we have $|k\alpha - h| < 1/n$ with $k \leq n$. So far we have proved, then, that, given any positive integer n, there exist integers k and h, with $n \geq k > 0$, such that

$$(4.2) \quad |k\alpha - h| < 1/n \quad \text{or} \quad |\alpha - h/k| < 1/nk.$$

The latter relation implies the inequality stated in the theorem, because $n \geq k$ implies that $1/nk \leq 1/k^2$.

To complete the proof of the theorem we must show that there are infinitely many such pairs of integers (h, k). Suppose, on the contrary, that there is only a finite number, say

$$(h_1, k_1), (h_2, k_2), \cdots, (h_r, k_r).$$

We prove that this supposition is false by finding another pair (h, k) satisfying (4.2). Define ϵ as the minimum of

$$|\alpha - h_1/k_1|, |\alpha - h_2/k_2|, \cdots, |\alpha - h_r/k_r|.$$

Since α is irrational, ϵ is positive. Choose n so that $1/n < \epsilon$, and then by the first part of the proof which led to (4.2) we can find a rational h/k so that

$$|\alpha - h/k| < 1/nk \leq 1/n < \epsilon.$$

By the definition of ϵ it follows that h/k is different from h_i/k_i for $i = 1, 2, \cdots, r$, and this establishes the theorem.

THEOREM 4.2. *Given any irrational α and any positive integer n, there exist integers h and k with $0 < k \leqq n$ such that $|k\alpha - h| < 1/n$.*

This result follows from the proof of Theorem 4.1, in particular the first relation (4.2).

THEOREM 4.3. *A real number α is irrational if and only if for any positive ϵ there exist infinitely many pairs of integers (x, y) such that $0 < |\alpha x - y| < \epsilon$.*

Proof. First, if α is irrational, we can get infinitely many pairs (x, y) from the pairs (k, h) of Theorem 4.1, using only those for which $1/k < \epsilon$. Conversely, if α is rational, say $\alpha = a/b$ with $b > 0$, then $|\alpha x - y|$ is a rational number for any pair of integers (x, y). Indeed $|\alpha x - y|$ is of the form u/b where u is some non-negative integer. Then, if we choose $\epsilon < 1/b$, we see that no pair of integers (x, y) can satisfy the inequality stated in the theorem. For, if (x, y) is a pair such that $|\alpha x - y| > 0$, then $|\alpha x - y| \geqq 1/b > \epsilon$.

2. A generalization. It is natural to inquire whether the results obtained above can be extended to the simultaneous approximation of irrationals: given α_1 and α_2, can we find k so that both $k\alpha_1$ and $k\alpha_2$ are arbitrarily close to integers? Again, can we find integers k_1 and k_2 so that the linear combination $k_1\alpha_1 + k_2\alpha_2$ is, in some specified manner, close to an integer? We shall generalize in both ways suggested by these questions.

We say that $(k_1, k_2, \cdots, k_m)$ is a *lattice point* in m-space if its coordinates are integers. Given any rectangular array or matrix of real numbers α_{ij} with $j = 1, 2, \cdots, m$ and $i = 1, 2, \cdots, n$, we set up the n linear forms $\sum_{j=1}^{m} \alpha_{ij}k_j$. We establish that the lattice point $(k_1, k_2, \cdots, k_m)$ can be chosen so that each of these n linear forms is arbitrarily close to an integer. To say this another way, if we regard the n linear forms as coordinates of a point in n-space, this point can be located arbitrarily close to some lattice point $(h_1, h_2, \cdots, h_n)$ in n-space by suitable choice of the integers k_j. This can be done in a trivial way by choosing each $k_j = 0$; so we require that not all the k_j vanish; that is we require that $\Sigma |k_j| \neq 0$. We now state the proposition formally.

THEOREM 4.4   *For positive integers m and n let α_{ij} be a collection of mn real numbers, with $i = 1, 2, \cdots, n$ and $j = 1, 2, \cdots, m$. Let $\tau \geqq 1$ be any real number, and define $T = -[-\tau]$, so that T is the smallest integer not less than τ. Then there exist lattice points $(k_1, k_2, \cdots, k_m)$ and $(h_1, h_2, \cdots, h_n)$ with $|k_j| \leqq T^{n/m}$ for all $j = 1, 2, \cdots, m$, and $\sum_{j=1}^{m} |k_j| \neq 0$, such that*

$$\left| \sum_{j=1}^{m} \alpha_{ij}k_j - h_i \right| < 1/\tau \quad \text{for} \quad i = 1, 2, \cdots, n.$$

Proof.   For any positive integer q there are $(q + 1)^m$ lattice points $(y_1, \cdots, y_m)$ with $0 \leqq y_j \leqq q$, and $(q + 1)^m$ corresponding points $(\omega_1, \cdots, \omega_n)$ defined by

$$\omega_i = \sum_{j=1}^{m} \alpha_{ij}y_j, \qquad i = 1, 2, \cdots, n.$$

Let x_i be the integer such that $0 \leqq x_i - \omega_i < 1$ for each $i = 1, 2, \cdots, n$; an alternative definition is $x_i = -[-\omega_i]$. Thus we obtain a collection Q of $(q + 1)^m$ points defined thus,

$$Q: \quad (x_1 - \omega_1, x_2 - \omega_2, \cdots, x_n - \omega_n),$$

which are points of the unit cube in n-dimensional space. The unit cube referred to here is in effect defined by the inequalities $0 \leqq x_i - \omega_i < 1$, so that it is a half-open cube. Divide this unit cube into T^n smaller cubes of side $1/T$ by means of parallel hyperplanes, each of these subcubes being half-open in the same way.

Next we set $q = [T^{n/m}]$ so that

$$(q + 1)^m = ([T^{n/m}] + 1)^m > (T^{n/m})^m = T^n.$$

Thus the $(q + 1)^m$ points of Q, being distributed in some fashion in the T^n subcubes of side $1/T$, cannot all lie in different subcubes. At least two points of Q lie in the same subcube, say the two points $(x_1 - \omega_1, \cdots, x_n - \omega_n)$ and $(x_1' - \omega_1', \cdots, x_n' - \omega_n')$, where $\omega_i' = \sum \alpha_{ij} y_j'$, and $(y_1', \cdots, y_m')$ is a lattice point distinct from $(y_1, \cdots, y_m)$ but with $0 \leqq y_j' \leqq q$. Hence we have

$$\frac{1}{T} > |x_i - \omega_i - x_i' + \omega_i'|$$
$$= \left| \sum_{j=1}^{m} \alpha_{ij}(y_j' - y_j) - (x_i' - x_i) \right|, \qquad i = 1, 2, \cdots, n.$$

These inequalities imply the conclusion of the theorem if we first write k_j for $y_j' - y_j$, h_i for $x_i' - x_i$, and then use the fact that $\tau < T$.

However, we must verify that the k_j satisfy the inequalities stated in the theorem. Since $0 \leqq y_j \leqq q$, and likewise for y_j', it follows that

$$|k_j| = |y_j' - y_j| \leqq q = [T^{n/m}] \leqq T^{n/m}.$$

Finally, since the lattice points $(y_1, \cdots, y_m)$ and $(y_1', \cdots, y_m')$ are distinct, we note that $k_j \neq 0$ for at least one j, and hence $\Sigma |k_j| \neq 0$.

COROLLARY 4.5. *Given any real numbers $\alpha_1, \alpha_2, \cdots, \alpha_m$ and any integer $t \geqq 1$, there exists a lattice point $(k_1, \cdots, k_m, h)$ with $|k_j| < t$ for all $j = 1, 2, \cdots, m$, and $\Sigma |k_j| \neq 0$,* *such that*

$$|k_1\alpha_1 + k_2\alpha_2 + \cdots + k_m\alpha_m - h| < 1/t^m.$$

Proof. In Theorem 4.4 we replace τ by t^m, n by 1, α_{1j} by α_j, and h_1 by h, so that $T = t^m$, $T^{n/m} = t$, and the result follows.

THEOREM 4.6.   *Given any real numbers* α_1, α_2, $\cdots$, α_n, *there exist infinitely many sets of integers* k, q_1, q_2, $\cdots$, q_n *with* $k > 0$ *such that*

$$(4.3) \quad |\alpha_i - q_i/k| < 1/(k\sqrt[n]{k}) \quad \text{for} \quad i = 1, 2, \cdots, n.$$

Proof. In Theorem 4.4 we replace m by 1, α_{i1} by α_i, and we employ τ only as a positive integer so that $\tau = T$. Thus we conclude that for any positive integer T there is a lattice point $(k_1, h_1, h_2, \cdots, h_n)$ with $0 < |k_1| \leq T^n$ such that

$$|\alpha_i k_1 - h_i| < 1/T \quad \text{for} \quad i = 1, 2, \cdots, n.$$

These inequalities also hold, obviously, for the lattice point $(-k_1, -h_1, -h_2, \cdots, -h_n)$; so select that one of this pair of lattice points having positive first member k_1 or $-k_1$, and designate it by $(k, q_1, q_2, \cdots, q_n)$. Thus we know that for any positive integer T there is a lattice point $(k, q_1, q_2, \cdots, q_n)$, with $0 < k < T^n$ such that

$$(4.4) \quad |\alpha_i k - q_i| < 1/T \quad \text{for} \quad i = 1, 2, \cdots, n.$$

This is equivalent to the inequalities (4.3) since we can divide through by the positive integer k and replace $1/T$ by the larger value $1/\sqrt[n]{k}$.

We must prove that there are infinitely many sets of integers satisfying (4.3). We separate the proof into two parts. First, if all of α_1, α_2, $\cdots$, α_n are rational, the whole result becomes trivial because we can ignore all the above theory and simply choose k as any common integral denominator of the α_i and then take $q_i = k\alpha_i$. And there are infinitely many choices for such a k.

Second, suppose that at least one of the α_i is irrational. There is no loss of generality in assuming that α_1 is irrational. If there were only a finite number of sets of integers k, q_1, $\cdots$, q_n satisfying (4.3), then there would be only a finite set of corresponding values $|\alpha_1 k - q_1|$, all positive since they are irrational. But each of these values would exceed $1/T$ provided we choose the positive integer T sufficiently large. This integer T would, by the procedure leading to (4.4), give us a different set of integers k, q_1, $\cdots$, q_n, and thus the proof of Theorem 4.6 is complete.

3. Linearly dependent sets. Theorem 4.6 can be strengthened in case the numbers α_1, $\cdots$, α_n are linearly dependent over the field of rational numbers, and this we now do.

THEOREM 4.7. *Assume that only m of the real numbers α_1, α_2, $\cdots$, α_n are linearly independent over the field of rational numbers. Then there exists a constant c and infinitely many sets of integers k, q_1, $\cdots$, q_n with $k > 0$ such that*

$$|\alpha_i - q_i/k| < c/(k\sqrt[m]{k}) \quad for \quad i = 1, 2, \cdots, n.$$

Proof. If $n = m$, the result is Theorem 4.6 with $c = 1$. Hence we may assume that $m < n$, and we arrange the notation so that α_1, $\cdots$, α_m are linearly independent, and each of α_{m+1}, $\cdots$, α_n can be represented as a linear combination of these with rational coefficients. Using a common denominator $b > 0$ for these rational coefficients, we have relations of the form, with integers a_{ij},

$$(4.5) \qquad \alpha_i = \frac{1}{b} \sum_{j=1}^{m} a_{ij}\alpha_j \quad for \quad i = m + 1, \cdots, n.$$

Applying Theorem 4.6 to $\alpha_1, \cdots, \alpha_m$, we see that there exist infinitely many sets of integers $k', q_1', \cdots, q_m'$ with $k' > 0$ for which

$$(4.6) \quad \left| \alpha_j - \frac{q_j'}{k'} \right| < \frac{1}{k' \sqrt[m]{k'}} \quad \text{for} \quad j = 1, 2, \cdots, m.$$

For $i > m$ we have, using (4.5) and (4.6),

$$(4.7) \quad \left| \alpha_i - \frac{1}{bk'} \sum_{j=1}^{m} a_{ij} q_j' \right| = \left| \frac{1}{b} \sum_{j=1}^{m} a_{ij} \alpha_j - \frac{1}{bk'} \sum_{j=1}^{m} a_{ij} q_j' \right|$$

$$\leqq \sum_{j=1}^{m} \left| \frac{a_{ij}}{b} \left(\alpha_j - \frac{q_j'}{k'} \right) \right| < \sum_{j=1}^{m} \left| \frac{a_{ij}}{b} \cdot \frac{1}{k' \sqrt[m]{k'}} \right|$$

$$= \frac{1}{bk' \sqrt[m]{k'}} \sum_{j=1}^{m} |a_{ij}|.$$

We define $k = bk'$, $q_i = bq_i'$ for $i = 1, \cdots, m$, and

$$q_i = \sum_{j=1}^{m} a_{ij} q_j' \quad \text{for} \quad i = m+1, \cdots, n,$$

so that there are infinitely many sets of integers $k, q_1, q_2, \cdots, q_n$. With this notation, (4.6) can be written as

$$\left| \alpha_j - \frac{q_j}{k} \right| < \frac{b \sqrt[m]{b}}{k \sqrt[m]{k}} \quad \text{for} \quad j = 1, \cdots, m,$$

and (4.7) as

$$\left| \alpha_i - \frac{q_i}{k} \right| < \frac{1}{k \sqrt[m]{k}} \cdot \sqrt[m]{b} \cdot \sum_{j=1}^{m} |a_{ij}|$$
$$\text{for} \quad i = m+1, \cdots, n.$$

These two inequalities establish the theorem if we define c as the maximum of the $n + 1 - m$ numbers

$$b\sqrt[m]{b}, \quad \sqrt[m]{b} \cdot \sum_{j=1}^{m} |a_{ij}|, \qquad i = m + 1, \cdots, n.$$

Notes on Chapter 4

All the arguments here are based on the pigeon-hole principle of Dirichlet: if there are $n + 1$ objects in n boxes, there must be at least one box containing two or more of the objects. Theorem 4.1 can be improved, and this is done in Chapter 6 (Theorem 6.1) by the use of continued fractions.

The results, both here and in Chapter 6, are only a small sample of the work that has been done with this topic; cf. J. F. Koksma, *Diophantische Approximationen*. An excellent exposition is to be found in G. H. Hardy and E. M. Wright, *Theory of Numbers*, Chapters 11, 23, 24.

CHAPTER 5

CONTINUED FRACTIONS

1. The Euclidean algorithm.   Consider a pair of integers u_0 and u_1, with $u_1 \neq 0$ and $(u_0, u_1) = 1$. The division algorithm shows that, if u_0 is divided by u_1, there is a unique quotient $[u_0/u_1]$ and a unique remainder, say u_2, with $0 \leq u_2 < u_1$. If $u_2 \neq 0$, the process continues with u_1 divided by u_2, and in this way we get the Euclidean algorithm

$$u_0 = u_1[u_0/u_1] + u_2$$

$$u_1 = u_2[u_1/u_2] + u_3$$

$$\cdot \quad \cdot \quad \cdot \quad \cdot \quad \cdot \quad \cdot \quad \cdot \quad \cdot \quad \cdot \quad \cdot \quad \cdot$$

(5.1) $$u_{i-1} = u_i[u_{i-1}/u_i] + u_{i+1}$$

$$\cdot \quad \cdot \quad \cdot \quad \cdot \quad \cdot \quad \cdot \quad \cdot \quad \cdot \quad \cdot \quad \cdot \quad \cdot$$

$$u_{m-1} = u_m[u_{m-1}/u_m] + u_{m+1}$$

$$u_m = u_{m+1}[u_m/u_{m+1}].$$

The remainders satisfy the inequality $0 < u_{i+1} < u_i$ for $1 \leq i \leq m$. The last non-zero remainder, u_{m+1}, is the g.c.d. of u_0 and u_1, and so by hypothesis $u_{m+1} = 1$. In the special case $u_1 = 1$, u_0/u_1 is an integer and $m = 0$.

If we write $\xi_i = u_i/u_{i+1}$ and $a_i = [\xi_i]$ for $0 \leq i \leq m$, we can put the equations (5.1) in the form

(5.2) $\quad \xi_{i-1} = a_{i-1} + 1/\xi_i, \qquad 1 \leq i \leq m; \qquad \xi_m = a_m.$

By eliminating ξ_1 from the cases $i = 1$ and $i = 2$ of equations (5.2), and then eliminating ξ_2 from the result and the case $i = 3$ of (5.2), and continuing this process until finally ξ_m is eliminated, we obtain

$$(5.3) \qquad \xi_0 = a_0 + \cfrac{1}{a_1 + \cfrac{}{\ddots \, + \cfrac{1}{a_{m-1} + \cfrac{1}{a_m}}}}$$

This is a continued fraction expansion of ξ_0. The integers a_i are called the *partial quotients*, since $a_i = [u_i/u_{i+1}]$ for $0 \leq i \leq m$. Now a_0 may be positive, zero, or negative, but $a_i \geq 1$ for $1 \leq i \leq m$, since $u_{i+1} < u_i$. Also we note that $a_m > 1$ if $m \geq 1$, because $u_{m+1} = 1$, $\xi_m = u_m = a_m$ and $u_m > u_{m+1}$.

Notation. For any real numbers $x_0, x_1, \cdots, x_n$ with $x_i > 0$ if $i > 0$, we shall write

$$(5.4) \qquad [x_0, x_1, \cdots, x_n] = x_0 + \cfrac{1}{x_1 + \cfrac{}{\ddots \, + \cfrac{1}{x_{n-1} + \cfrac{1}{x_n}}}}$$

This finite continued fraction is said to be *simple* in case the x_i are integers, all positive except perhaps x_0. For example, the right side of (5.3) is a finite simple continued fraction, and so it could be represented by the notation $[a_0, a_1, \cdots, a_m]$.

We shall make free use of the formulas

$$[x_0, x_1, \cdots, x_n] = x_0 + \cfrac{1}{[x_1, x_2, \cdots, x_n]}$$

$$= \left[x_0, x_1, \cdots, x_{n-2}, x_{n-1} + \frac{1}{x_n} \right].$$

This notation (5.4) for a continued fraction is in very minor conflict with the greatest integer notation, as follows. Setting $n = 0$ in (5.4), we have $[x_0] = x_0$, and this is inconsistent with the greatest integer notation whenever x_0 is not an integer. However, this will cause no difficulty because our primary concern is with *simple* continued fractions. In any case, the meaning will be clear by the context.

2. Uniqueness. The simple continued fraction expansion (5.3) of ξ_0 is not unique, because

$$(5.5) \qquad \xi_0 = [a_0, a_1, \cdots, a_{m-1}, a_m]$$

$$= [a_0, a_1, \cdots, a_{m-1}, a_m - 1, 1].$$

Note that this equation is correct even if ξ_0 is an integer, in which case $m = 0$ and $a_0 = \xi_0$. We now establish that these are the only simple continued fraction expansions of any rational number ξ_0.

LEMMA 5.1. *If two finite simple continued fractions are equal, say $[a_0, a_1, \cdots, a_m] = [b_0, b_1, \cdots, b_n]$, and if $a_m > 1$ and $b_n > 1$, then $m = n$ and $a_i = b_i$ for $0 \leqq i \leqq n$.*

Proof. Writing y_i for $[b_i, b_{i+1}, \cdots, b_n]$, we see that

$$(5.6) \quad y_{i-1} = [b_{i-1}, b_i, \cdots, b_n]$$

$$= b_{i-1} + \cfrac{1}{[b_i, b_{i+1}, \cdots, b_n]} = b_{i-1} + \frac{1}{y_i}.$$

Hence we have $y_{i-1} > b_{i-1}$ and $y_{i-1} > 1$ for $2 \leqq i \leqq n$, and $y_n = b_n > 1$. It follows that $b_i = [y_i]$ for $0 \leqq i \leqq n$.

Now if we identify $y_0 = [b_0, \cdots, b_n]$ with ξ_0, we can compare equations (5.6) and (5.2). First we have that $a_0 = [\xi_0] = [y_0] = b_0$. Next we obtain

$$\frac{1}{\xi_1} = \xi_0 - a_0 = y_0 - b_0 = \frac{1}{y_1},$$

$$\xi_1 = y_1, \qquad a_1 = [\xi_1] = [y_1] = b_1.$$

This process continues by mathematical induction: from $\xi_{i-1} = y_{i-1}$ and $a_{i-1} = b_{i-1}$ it follows that

$$\frac{1}{\xi_i} = \xi_{i-1} - a_{i-1} = y_{i-1} - b_{i-1} = \frac{1}{y_i},$$

$$\xi_i = y_i, \qquad a_i = [\xi_i] = [y_i] = b_i.$$

Furthermore we have that $m = n$. For suppose that $m < n$, say. The induction process would bring us to the results $\xi_m = y_m$, $a_m = b_m$, and these would stand in contradiction to $\xi_m = a_m$ and $y_m > b_m$ from (5.2) and (5.6). The case $m > n$ leads to a contradiction in a symmetrical way, and so the lemma is proved.

THEOREM 5.2. *Any finite simple continued fraction represents a rational number. Conversely, any rational number ξ_0 can be expanded in a finite simple continued fraction in exactly two ways.*

Proof. The first remark is obvious. Concerning the converse, the "two ways" are given by equation (5.5). By using this equation to eliminate 1 as a last partial quotient, we then apply Lemma 5.1 to establish the theorem.

3. Infinite continued fractions. Let $b_0, b_1, b_2, \cdots$ be an infinite sequence of integers such that $b_i \geqq 1$ for $i \geqq 1$. Define the integers h_n and k_n by the equations

$$h_{-2} = 0, \quad h_{-1} = 1, \quad h_i = b_i h_{i-1} + h_{i-2} \quad \text{for} \quad i \geqq 0,$$

(5.7)

$$k_{-2} = 1, \quad k_{-1} = 0, \quad k_i = b_i k_{i-1} + k_{i-2} \quad \text{for} \quad i \geqq 0.$$

It is clear that $1 = k_0 \leqq k_1 < k_2 < \cdots$.

LEMMA 5.3.   *If x is any positive real number, then*

$$[b_0, b_1, \cdots, b_{n-1}, x] = \frac{x h_{n-1} + h_{n-2}}{x k_{n-1} + k_{n-2}}.$$

Proof.   We use induction. For $n = 0$ the result reduces to $[x] = x$ by use of equations (5.7), where $[x]$ here means the continued fraction with only one partial quotient. If $n = 1$, we have

$$[b_0, x] = b_0 + \frac{1}{x} = \frac{x b_0 + 1}{x} = \frac{x h_0 + h_{-1}}{x k_0 + k_{-1}}.$$

Assuming the result for $[b_0, \cdots, b_{n-1}, x]$, we can write

$$[b_0, \cdots, b_{n-1}, b_n, x] = \left[b_0, \cdots, b_{n-1}, b_n + \frac{1}{x} \right]$$

$$= \frac{\left(b_n + \dfrac{1}{x} \right) h_{n-1} + h_{n-2}}{\left(b_n + \dfrac{1}{x} \right) k_{n-1} + k_{n-2}}$$

$$= \frac{x(b_n h_{n-1} + h_{n-2}) + h_{n-1}}{x(b_n k_{n-1} + k_{n-2}) + k_{n-1}}$$

$$= \frac{x h_n + h_{n-1}}{x k_n + k_{n-1}}.$$

LEMMA 5.4. *If we define $z_n = [b_0, \ b_1, \ \cdots, \ b_n]$ for $n = 0, 1, 2, \cdots$, then $z_n = h_n/k_n$.*

Proof. Replace x by b_n in Lemma 5.3, and apply (5.7).

LEMMA 5.5 *For $i \geqq 1$ we have*

$$h_i k_{i-1} - h_{i-1} k_i = (-1)^{i-1} \quad and \quad z_i - z_{i-1} = \frac{(-1)^{i-1}}{k_i k_{i-1}}.$$

For $i \geqq 2$ we have

$$h_i k_{i-2} - h_{i-2} k_i = (-1)^i b_i \quad and \quad z_i - z_{i-2} = \frac{(-1)^i b_i}{k_i k_{i-2}}.$$

Also the fraction h_i/k_i is in lowest terms.

Proof. From (5.7) we see that $h_{-1} k_{-2} - h_{-2} k_{-1} = 1$, and by mathematical induction

$$h_i k_{i-1} - h_{i-1} k_i$$
$$= (b_i h_{i-1} + h_{i-2}) k_{i-1} - h_{i-1}(b_i k_{i-1} + k_{i-2})$$
$$= -(h_{i-1} k_{i-2} - h_{i-2} k_{i-1}) = (-1)^{i-1}.$$

Thus the first result of the lemma holds even for $i = -1$ and $i = 0$. The second result, the formula for $z_i - z_{i-1}$, is obtained by division by $k_i k_{i-1}$, and here the restriction $i \geqq 1$ is needed since $k_{-1} = 0$.

Analogously we can write $h_0 k_{-2} + h_{-2} k_0 = b_0$, and by induction

$$h_i k_{i-2} - h_{i-2} k_i$$
$$= (b_i h_{i-1} + h_{i-2}) k_{i-2} - h_{i-2}(b_i k_{i-1} + k_{i-2})$$
$$= b_i(h_{i-1} k_{i-2} - h_{i-2} k_{i-1}) = (-1)^i b_i.$$

Division by $k_i k_{i-2}$ will then yield the final formula of the lemma. Finally, it is clear that h_i/k_i is in lowest terms, since any divisor of h_i and k_i is a divisor of $(-1)^{i-1}$.

LEMMA 5.6. *The values z_i satisfy the inequalities*

$$z_0 < z_2 < z_4 < \cdots < z_5 < z_3 < z_1.$$

Furthermore $\lim z_n$ *exists, and* $z_{2i} < \lim z_n < z_{2i+1}$ *for any non-negative integer* i.

Proof. Recalling that k_i is positive for $i \geqq 0$, and b_i is positive for $i \geqq 1$, we see by Lemma 5.5 that $z_{2i} < z_{2i+2}$, $z_{2i-1} > z_{2i+1}$, and $z_{2i} < z_{2i-1}$. Thus the monotonically increasing sequence z_0, z_2, z_4, $\cdots$ is bounded above by z_1, and so tends to a limit. Likewise the monotonically decreasing sequence z_1, z_3, z_5, $\cdots$ is bounded below by z_0, and so tends to a limit. These limits are equal because Lemma 5.5 also implies that $z_i - z_{i-1}$ tends to zero with increasing i, since the integers k_i are increasing.

The chain of results which we have obtained in this section suggests the following definition of an infinite simple continued fraction and its value.

DEFINITION. Let b_0, b_1, b_2, $\cdots$ be an infinite sequence of integers such that $b_i \geqq 1$ for $i \geqq 1$. The limit of the finite simple continued fraction $[b_0, b_1, \cdots, b_n]$ as n increases indefinitely is called an *infinite simple continued fraction*, with the notation $[b_0, b_1, b_2, \cdots]$.

Thus we have

$$(5.8) \qquad [b_0, b_1, b_2, \cdots] = \lim_{n \to \infty} [b_0, b_1, \cdots, b_n]$$

$$= \lim_{n \to \infty} \frac{h_n}{k_n} = \lim_{n \to \infty} z_n,$$

where h_n, k_n, and z_n are defined in (5.7) and Lemma 5.4. The rational number h_n/k_n, or in alternative notation z_n or $[b_0, b_1, \cdots, b_n]$, is called the *n-th convergent* to the infinite continued fraction.

THEOREM 5.7. *Any infinite simple continued fraction* $[b_0, b_1, b_2, \cdots]$ *is an irrational number.*

Proof. Writing θ for $[b_0, b_1, b_2, \cdots]$, we see by Lemma 5.6 that θ lies between consecutive convergents z_n and

z_{n+1}: that is, between h_n/k_n and h_{n+1}/k_{n+1}. Thus we have, by Lemma 5.5,

$$0 < |k_n\theta - h_n| = k_n|\theta - z_n| < k_n|z_{n+1} - z_n|$$

$$= k_n \frac{1}{k_n k_{n+1}} = \frac{1}{k_{n+1}}.$$

Once again we make use of the fact that the k_n are positive integers, increasing with n. Given any positive ϵ, we see that we can, by taking all sufficiently large n, find infinitely many pairs h, k such that $0 < |k\theta - h| < \epsilon$. Hence θ is irrational by Theorem 4.3. (This is the only place in the present chapter where a proof depends on a result from an earlier chapter. However, it may be noted that the dependence is on the trivial part of Theorem 4.3, the proof of which follows the statement of that result.)

We now establish that two different infinite simple continued fractions cannot have the same value.

LEMMA 5.8.   *Let $\theta = [b_0, b_1, b_2, \cdots]$.   Then $b_0 = [\theta]$, where by $[\theta]$ is meant the greatest integer not exceeding θ.*

Proof. By Lemma 5.6 we have $z_0 < \theta < z_1$, which is the same as $b_0 < \theta < b_0 + 1/b_1$. But $b_1 \geqq 1$, so that $b_0 < \theta < b_0 + 1$, and this completes the proof.

LEMMA 5.9.   *With θ as in the previous lemma, and $\theta_1 = [b_1, b_2, \cdots]$, we have $\theta = b_0 + 1/\theta_1$.*

Proof. Note that $\theta_1 \neq 0$ by Theorem 5.7. We can write

$$\theta = \lim_{n \to \infty} [b_0, b_1, \cdots, b_n] = \lim_{n \to \infty} \left\{ b_0 + \frac{1}{[b_1, \cdots, b_n]} \right\}$$

$$= b_0 + \frac{1}{\lim [b_1, \cdots, b_n]} = b_0 + \frac{1}{\theta_1}.$$

THEOREM 5.10.   *Two distinct infinite simple continued fractions represent different irrational numbers.*

Proof. Suppose that $[b_0, b_1, b_2, \cdots] = [a_0, a_1, a_2, \cdots]$ $= \theta$. Then Lemma 5.8 states that $[\theta] = b_0 = a_0$. Next we apply Lemma 5.9 to show that $[b_1, b_2, \cdots] = [a_1, a_2, \cdots]$. This process continues by mathematical induction, so that $a_n = b_n$ for all n.

4. Infinite continued fraction expansions. The converse of Theorem 5.7 is also true: any irrational number can be expanded into an infinite simple continued fraction. To establish this, let ξ_0 be any irrational number. Then define the sequence of integers $a_0, a_1, a_2, \cdots$, and the sequence of irrationals $\xi_1, \xi_2, \xi_3, \cdots$ by the equations

$$a_0 = [\xi_0], \qquad \xi_1 = 1/(\xi_0 - a_0)$$

(5.9)
$$a_1 = [\xi_1], \qquad \xi_2 = 1/(\xi_1 - a_1)$$

$$\cdots \cdots \cdots \cdots \cdots \cdots$$

$$a_i = [\xi_i], \qquad \xi_{i+1} = 1/(\xi_i - a_i).$$

The relation $\xi_{i-1} - a_{i-1} = \xi_{i-1} - [\xi_{i-1}]$, together with the fact that ξ_{i-1} is irrational, indicates clearly that $0 < \xi_{i-1} - a_{i-1} < 1$, whence $\xi_i = (\xi_{i-1} - a_{i-1})^{-1} > 1$ and $a_i = [\xi_i] \geqq 1$ for all $i \geqq 1$.

Making use of (5.9) in the form $\xi_{i-1} = a_{i-1} + 1/\xi_i$, we see that

$$\xi_0 = a_0 + \frac{1}{\xi_1} = a_0 + \frac{1}{a_1 + 1/\xi_2} = [a_0, a_1, \xi_2].$$

Continuation of this process by mathematical induction gives $\xi_0 = [a_0, a_1, a_2, \cdots, a_{n-1}, \xi_n]$. We now identify the a_i with the b_i of the previous section; that is, we write $a_0 = b_0$, $a_1 = b_1$, etc. Then Lemma 5.3, with x replaced by ξ_n, states that

$$(5.10) \quad \xi_0 = [a_0, a_1, \cdots, a_{n-1}, \xi_n] = \frac{\xi_n h_{n-1} + h_{n-2}}{\xi_n k_{n-1} + k_{n-2}}.$$

The values h_i and k_i satisfy recurrence relations (5.7), which can be rewritten as

$$(5.11) \quad h_i = a_i h_{i-1} + h_{i-2}, \qquad k_i = a_i k_{i-1} + k_{i-2}.$$

Thus we have, using Lemma 5.5,

$$(5.12) \quad \xi_0 - z_{n-1} = \xi_0 - \frac{h_{n-1}}{k_{n-1}} = \frac{\xi_n h_{n-1} + h_{n-2}}{\xi_n k_{n-1} + k_{n-2}} - \frac{h_{n-1}}{k_{n-1}}$$

$$= \frac{-(h_{n-1}k_{n-2} - h_{n-2}k_{n-1})}{k_{n-1}(\xi_n k_{n-1} + k_{n-2})}$$

$$= \frac{(-1)^{n-1}}{k_{n-1}(\xi_n k_{n-1} + k_{n-2})}.$$

Hence $\xi_0 - z_{n-1}$ is positive for odd values of n, and negative for even values; that is

$$(5.13) \quad z_{2n} = h_{2n}/k_{2n} < \xi_0 < h_{2n-1}/k_{2n-1} = z_{2n-1}.$$

This implies that

$$\xi_0 = \lim_{n \to \infty} z_n = [a_0, a_1, a_2, \cdots],$$

so that equations (5.9) constitute an algorithm for determining an infinite simple continued fraction expansion of any irrational ξ_0. Uniqueness has been established in Theorem 5.10; so we can summarize as follows.

THEOREM 5.11. *Every irrational number ξ_0 has a unique representation as an infinite simple continued fraction $[a_0, a_1, a_2, \cdots]$, and conversely. The integers a_i are positive for $i \geqq 1$. The n-th convergent, h_n/k_n, is the finite continued fraction $[a_0, a_1, \cdots, a_n]$. The denominators k_n form a monotonically increasing sequence of integers, for $n \geqq 1$. The even convergents, i.e., h_n/k_n for $n = 0, 2, 4, \cdots$, are monotonically increasing with ξ_0 as a limit; the odd convergents are monotonically decreasing with ξ_0 as a limit.*

5. The convergents as approximations. It will be convenient in the sequel often to drop the subscript from ξ_0, and write simply ξ for the irrational number with continued fraction expansion $[a_0, a_1, a_2, \cdots]$ and convergents h_n/k_n.

LEMMA 5.12.   *For any $n \geq 0$, we have*

$$|\xi - h_n/k_n| < 1/k_nk_{n+1}.$$

Proof.   We use (5.12), (5.9), and (5.11) to write

$$\left| \xi - \frac{h_n}{k_n} \right| = \frac{1}{k_n(\xi_{n+1}k_n + k_{n-1})} < \frac{1}{k_n(a_{n+1}k_n + k_{n-1})}$$

$$= \frac{1}{k_nk_{n+1}}.$$

Next we establish that the convergents form a chain of consecutively better approximations to an irrational number.

THEOREM 5.13.   *The convergents h_n/k_n of the continued fraction expansion of an irrational number ξ satisfy, for $n \geq 1$,*

$$\left| \xi - \frac{h_n}{k_n} \right| < \left| \xi - \frac{h_{n-1}}{k_{n-1}} \right|$$

and

$$|\xi k_n - h_n| < |\xi k_{n-1} - h_{n-1}|.$$

Proof.   We establish the second inequality, whence the first is true a fortiori since the positive integers k_n increase with n, except for the possibility $k_0 = k_1$. By (5.9) we note that $a_n + 1 > \xi_n$. Hence by (5.11) we have

$$\xi_nk_{n-1} + k_{n-2} < (a_n + 1)k_{n-1} + k_{n-2}$$

$$= k_n + k_{n-1} \leq a_{n+1}k_n + k_{n-1} = k_{n+1},$$

for all $n \geq 1$. This with (5.12) implies that

$$\left| \xi - \frac{h_{n-1}}{k_{n-1}} \right| = \frac{1}{k_{n-1}(\xi_n k_{n-1} + k_{n-2})} > \frac{1}{k_{n-1}k_{n+1}}.$$

Multiplication by k_{n-1} gives $|k_{n-1}\xi - h_{n-1}| > k_{n+1}^{-1}$ for all $n \geq 1$. But Lemma 5.12 says that $|k_n\xi - h_n| < k_{n+1}^{-1}$. These two inequalities establish the lemma.

The convergents h_n/k_n are the best approximations to an irrational ξ in the following sense.

THEOREM 5.14. *If there is a rational number a/b with b positive such that $|\xi - a/b| < |\xi - h_n/k_n|$ for some $n > 0$, then $b > k_n$. Indeed, if there is a rational number a/b with b positive such that $|b\xi - a| < |k_n\xi - h_n|$ for some $n > 0$, then $b > k_n$.*

Proof. The second assertion implies the first, as the following argument shows. Assume that $|\xi - a/b| < |\xi - h_n/k_n|$ and that $b \leq k_n$. Then the product of these inequalities gives $|b\xi - a| < |k_n\xi - h_n|$, and by the second part of the theorem it would follow that $b > k_n$, a contradiction.

In order to prove the second part of the theorem, we assume that $|b\xi - a| < |k_n\xi - h_n|$ and that $b \leq k_n$. Hence $b < k_{n+1}$, since $n > 0$. Integers x and y exist such that

$$b = xk_n + yk_{n+1}, \qquad a = xh_n + yh_{n+1},$$

because the determinant of coefficients is ± 1 by Lemma 5.5. Next we argue that neither x nor y vanishes. First, if $y = 0$, then $a = xh_n$, $b = xk_n$, $x \neq 0$, and

$$|b\xi - a| = |x| \cdot |k_n\xi - h_n| \geq |k_n\xi - h_n|,$$

contrary to hypothesis. Second, if $x = 0$, then $y \neq 0$ and $b = yk_{n+1}$, which contradicts $b < k_{n+1}$.

Furthermore, x and y are of opposite signs. If $y < 0$, then the equation $xk_n = b - yk_{n+1}$ shows that $x > 0$. If $y > 0$, then $b < yk_{n+1}$ so that $xk_n < 0$ and $x < 0$. By (5.13)

$$k_n\xi - h_n \quad \text{and} \quad k_{n+1}\xi - h_{n+1}$$

have opposite signs, and so

$$x(k_n\xi - h_n) \quad \text{and} \quad y(k_{n+1}\xi - h_{n+1})$$

have the same sign. It is also clear that

$$b\xi - a = x(k_n\xi - h_n) + y(k_{n+1}\xi - h_{n+1}).$$

Taking absolute values, we get

$$\begin{aligned}
|b\xi - a| &= |x(k_n\xi - h_n)| + |y(k_{n+1}\xi - h_{n+1})| \\
&> |x(k_n\xi - h_n)| \geqq |k_n\xi - h_n|.
\end{aligned}$$

Thus we have a contradiction, and the theorem is proved.

The assumption $n > 0$ is needed in the theorem. Consider the case $\xi = \sqrt{3}$, $a = 2$, $b = 1$. Then $h_0/k_0 = 1/1$, and

$$|\xi - a/b| = |\sqrt{3} - 2| < |\sqrt{3} - 1| = |\xi - h_0/k_0|,$$

whereas $b = 1 = k_0$.

6. Periodic continued fractions. We say that an infinite simple continued fraction $[a_0, a_1, a_2, \cdots]$ is *periodic* if there exist integers n and s such that $a_r = a_{n+r}$ for all $r > s$. For example, $[2, 3, 4, 5, 4, 5, 4, 5, \cdots]$ is periodic. To evaluate such a continued fraction, write θ for the purely repeating part, $\theta = [4, 5, 4, 5, \cdots] = [4, 5, \theta]$. Hence we have $\theta = 4 + (5 + \theta^{-1})^{-1}$, so that $\theta = (10 + 2\sqrt{30})/5$. Then the original fraction is $[2, 3, 4, 5, 4, 5, \cdots] = [2, 3, \theta] = (7\theta + 2)/(3\theta + 1)$, which is readily seen to be a quadratic irrational. Another example is $\alpha = [1, 1, 1, \cdots] = (1 + \sqrt{5})/2$, a value that can be found by means of the equation $\alpha = [1, \alpha] = 1 + \alpha^{-1}$.

Any periodic continued fraction is quadratic, as we now show by an extension of the above argument. Let us use the notation

$$\xi = [b_0, b_1, \cdots, b_s, a_0, a_1, \cdots, a_{n-1}, a_0, a_1, \cdots, a_{n-1}, \cdots],$$

and write

$$\theta = [a_0, a_1, \cdots, a_{n-1}, a_0, a_1, \cdots, a_{n-1}, \cdots]$$

$$= [a_0, a_1, \cdots, a_{n-1}, \theta].$$

Then formula (5.10) gives

$$\theta = \frac{\theta h_{n-1} + h_{n-2}}{\theta k_{n-1} + k_{n-2}}.$$

Thus θ satisfies a quadratic equation with integral coefficients, so that θ is either rational or a quadratic irrational. But θ is an infinite continued fraction and so is irrational, and hence a quadratic irrational. Returning to ξ, we have

$$\xi = [b_0, b_1, \cdots, b_s, \theta] = \frac{p\theta + p'}{q\theta + q'},$$

where p/q and p'/q' are the last two convergents to $[b_0, b_1, \cdots, b_s]$. Again we can argue from this last equation that ξ is either rational or a quadratic irrational, and the former is ruled out because ξ is an infinite continued fraction.

We have proved the first part of the following result.

THEOREM 5.15. *Any periodic simple continued fraction is a quadratic irrational, and conversely.*

Proof. The first part of the proof preceded the statement of the theorem; so now we consider a quadratic irrational $\xi = \xi_0 = (a + \sqrt{b})/c$. Here b is a positive integer, not a perfect square. The integers a and c may be positive or negative, and so the form given to ξ_0 is perfectly general. Next, ξ_0 may be written in the form

$$\xi_0 = \frac{ac + \sqrt{bc^2}}{c^2} \quad \text{if} \quad c > 0, \quad \text{or}$$

$$\xi_0 = \frac{-ac + \sqrt{bc^2}}{-c^2} \quad \text{if} \quad c < 0.$$

Noting that c^2 is a divisor of $bc^2 - (ac)^2$, we may write

$$\xi_0 = \frac{P_0 + \sqrt{D}}{Q_0}, \quad \text{with} \quad Q_0 \mid (D - P_0^2),$$

where P_0, Q_0, and D are integers, $D > 0$, D not a perfect square.

We now establish that the infinite simple continued fraction expansion of ξ_0 is $[a_0, a_1, a_2, \cdots]$, where the a_i are defined by the following recursion relations:

$$(5.14) \qquad a_i = [\xi_i], \qquad \xi_i = \frac{P_i + \sqrt{D}}{Q_i},$$

$$P_{i+1} = a_i Q_i - P_i, \qquad Q_{i+1} = \frac{D - P_{i+1}^2}{Q_i}.$$

First we note that P_i and Q_i are integers for all $i \geqq 0$. To prove this, we use induction, assuming that P_i and Q_i are integers such that $Q_i \mid (D - P_i^2)$. Then P_{i+1} is an integer, and the equation

$$Q_{i+1} = \frac{D - P_{i+1}^2}{Q_i} = \frac{D - P_i^2 + 2a_i Q_i P_i - a_i^2 Q_i^2}{Q_i}$$

shows that Q_{i+1} is an integer. Furthermore, we have

$Q_i = (D - P_{i+1}^2)/Q_{i+1}$, so that $Q_{i+1}|(D - P_{i+1}^2)$, and the induction is complete.   Next we note that

$$\xi_i = \frac{P_i + \sqrt{D}}{Q_i} = a_i + \frac{P_i - a_iQ_i + \sqrt{D}}{Q_i}$$

$$= a_i + \frac{\sqrt{D} - P_{i+1}}{Q_i} = a_i + \frac{D - P_{i+1}^2}{Q_i(\sqrt{D} + P_{i+1})}$$

$$= a_i + \frac{1}{\dfrac{\sqrt{D} + P_{i+1}}{Q_{i+1}}} = a_i + \frac{1}{\xi_{i+1}}\cdot$$

This equality together with $a_i = [\xi_i]$ proves by (5.9) that

$$\xi_0 = [a_0, a_1, a_2, \cdots] = [a_0, a_1, \cdots, a_{n-1}, \xi_n].$$

By ξ_0' and ξ_n' we shall mean the conjugates of ξ_0 and ξ_n: that is, what is obtained from these when $\sqrt{D}$ is replaced by $-\sqrt{D}$. Taking conjugates in (5.10), we have

$$\xi_0' = \frac{\xi_n' h_{n-1} + h_{n-2}}{\xi_n' k_{n-1} + k_{n-2}}\cdot$$

Solving for ξ_n', we get

$$\xi_n' = -\frac{k_{n-2}}{k_{n-1}}\left(\frac{\xi_0' - h_{n-2}/k_{n-2}}{\xi_0' - h_{n-1}/k_{n-1}}\right).$$

Now $\lim h_n/k_n = \xi_0$, so that the term in parentheses tends to 1 as n tends to infinity. Hence for n sufficiently large, say $n \geqq N$, the term in parentheses is positive and ξ_n' is negative. But $\xi_n > 0$ for $n > 0$, and so $\xi_n - \xi_n' > 0$ for $n \geqq N$. By (5.14) this says that $2\sqrt{D}/Q_n > 0$, and so $Q_n > 0$ for $n \geqq N$. But by (5.14) we have $Q_nQ_{n+1} + P_{n+1}^2 = D$. Hence, for $n \geqq N$ we conclude that $Q_n < D$,

$|P_{n+1}| < \sqrt{D}$. Thus the P_n and Q_n range over a finite set of values, and so there exist distinct subscripts j and k so that $P_j = P_k$ and $Q_j = Q_k$. Hence $\xi_j = \xi_k$, and the continued fraction expansion $[a_0, a_1, \cdots]$ is periodic.

Notes on Chapter 5

The original draft of this chapter was revised extensively, particularly the order of the presentation and development of proofs, along lines suggested by Professor H. S. Zuckerman. The proof of Theorem 5.15 is Zuckerman's. We have included a little more material than is needed for the next chapter, in order to give some measure of completeness to the treatment. The classic work on continued fractions is O. Perron, *Die Lehre von den Kettenbrüchen*, Leipzig, Teubner (1929). An interesting introduction to the subject is given in Chapter 10 of Hardy and Wright, and a more elementary formulation in Chapter 4 of H. Davenport, *The Higher Arithmetic*.

CHAPTER 6

FURTHER
DIOPHANTINE APPROXIMATIONS

1. A basic result. This section is devoted to an improvement of Theorem 4.1, and this improvement is shown to be best possible in the next section. The last part of this chapter deals with the distribution of the fractional parts of the positive integral multiples of an irrational number.

THEOREM 6.1. *For any irrational ξ there exist infinitely many rational numbers h/k such that*

$$\left| \xi - \frac{h}{k} \right| < \frac{1}{\sqrt{5}k^2}.$$

Proof. We make use of the continued fraction expansion $[a_0, a_1, a_2, \cdots]$ of ξ, and the convergents h_n/k_n thereto. The proof consists in establishing that, of any three consecutive convergents, at least one satisfies the inequality of the theorem. (Actually the proof is not valid for the *first* three convergents, h_0/k_0, h_1/k_1, and h_2/k_2, one of which satisfies the inequality nevertheless.) Write c_{n+1} for k_{n-1}/k_n, so that by (5.12)

$$\left| \xi - \frac{h_n}{k_n} \right| = \frac{1}{k_n(\xi_{n+1}k_n + k_{n-1})} = \frac{1}{k_n^2(\xi_{n+1} + c_{n+1})}.$$

A comparison of this with the inequality of the theorem indicates that the question turns around the relative sizes of $\xi_{n+1} + c_{n+1}$ and $\sqrt{5}$. We shall assume that $\xi_i + c_i \leqq \sqrt{5}$ for three consecutive values of i, namely $i = n - 1$, n, $n + 1$, and from this will arise a contradiction, which will establish the theorem.

First we note from (5.11) that

$$(6.1) \quad \frac{1}{c_i} = \frac{k_{i-1}}{k_{i-2}} = \frac{a_{i-1}k_{i-2} + k_{i-3}}{k_{i-2}} = a_{i-1} + c_{i-1}.$$

Also from (5.9) we recall that $\xi_{n-1} = a_{n-1} + 1/\xi_n$, and we eliminate a_{n-1} from this and from formula (6.1) in the case $i = n$, to obtain

$$\frac{1}{c_n} + \frac{1}{\xi_n} = c_{n-1} + \xi_{n-1} \leqq \sqrt{5}, \qquad \frac{1}{\xi_n} \leqq \sqrt{5} - \frac{1}{c_n}.$$

But $\xi_n \leqq \sqrt{5} - c_n$ by hypothesis, and, since $\xi_n > 0$, we can multiply these two inequalities to get

$$1 \leqq 6 - \sqrt{5}\left(c_n + \frac{1}{c_n}\right), \qquad c_n + \frac{1}{c_n} \leqq \sqrt{5}.$$

The equality sign may be deleted here since c_n is rational. Also c_n is positive, and we use it as a multiplier to arrive at

$$c_n^2 + 1 - \sqrt{5}\,c_n < 0, \qquad \left(\frac{\sqrt{5}}{2} - c_n\right)^2 < \frac{1}{4}.$$

Furthermore $c_n < 1$ by Theorem 5.11, whence we can take square roots,

$$\frac{\sqrt{5}}{2} - c_n < \frac{1}{2}, \qquad c_n > \frac{1}{2}(\sqrt{5} - 1).$$

We have obtained this result by use of the cases $i = n - 1$ and $i = n$ of the inequality $\xi_i + c_i \leqq \sqrt{5}$. The latter inequality was assumed also for the case $i = n + 1$, and so we can also draw the conclusion that

$$c_{n+1} > \tfrac{1}{2}(\sqrt{5} - 1), \qquad c_{n+1}^{-1} < \tfrac{1}{2}(\sqrt{5} + 1).$$

Using these results together with formula (6.1) in the case $i = n + 1$, we obtain

$$a_n = \frac{1}{c_{n+1}} - c_n < \frac{1}{2}(\sqrt{5} + 1) - \frac{1}{2}(\sqrt{5} - 1) = 1,$$

which contradicts Theorem 5.11.

Three consecutive convergents were used in the proof, and this number cannot be reduced. For example, if $\xi = (4 - \sqrt{2})^{-1}$, then $a_0 = 0$, $a_1 = 2$, $a_2 = 1$, and the convergents h_1/k_1 and h_2/k_2, with values $\tfrac{1}{2}$ and $\tfrac{1}{3}$, fail to satisfy the inequality of Theorem 6.1.

2. Best possible approximations. The constant $\sqrt{5}$ in Theorem 6.1 is best possible in the sense of the following result.

THEOREM 6.2.   *The constant $\sqrt{5}$ in Theorem 6.1 cannot be replaced by any larger value.*

Proof. There are values of ξ for which the constant $\sqrt{5}$ in Theorem 6.1 can be improved, but $\xi = \tfrac{1}{2}(\sqrt{5} + 1)$ is not such a value, and this is the irrational which we use to establish Theorem 6.2.

Assume, then, that there are infinitely many rationals h/k for which

$$\left| \frac{\sqrt{5} + 1}{2} - \frac{h}{k} \right| < \frac{1}{ck^2},$$

where c is some constant exceeding $\sqrt{5}$. Define x, a function of h and k, by the equation

$$\frac{\sqrt{5} + 1}{2} - \frac{h}{k} = \frac{1}{xk^2},$$

so that $|x| > c > \sqrt{5}$. Then we can write

$$\frac{1}{xk} - \frac{\sqrt{5}\,k}{2} = \frac{k}{2} - h,$$

and, squaring and simplifying, we have

$$\frac{1}{x^2 k^2} - \frac{\sqrt{5}}{x} = h^2 - hk - k^2.$$

The right side of this equation is an integer, and so must the left side be. But the left side has absolute value

$$\left| \frac{1}{x^2 k^2} - \frac{\sqrt{5}}{x} \right| \leq \left| \frac{1}{x^2 k^2} \right| + \left| \frac{\sqrt{5}}{x} \right| < \frac{1}{k^2} + \frac{\sqrt{5}}{c}.$$

Now $\sqrt{5}/c < 1$; so we can choose k sufficiently large that $1/k^2 + \sqrt{5}/c < 1$. For such a value of k, and the corresponding h, it would follow that

$$h^2 - hk - k^2 = 0.$$

We may presume that h/k is in lowest terms, and it follows that the last equation is impossible in positive integers since it implies that any prime factor of k is a prime factor of h.

3. Uniform distributions. Let S be a sequence of numbers, or points, $\alpha_1, \alpha_2, \alpha_3, \cdots$, in the unit interval, so that $0 \leq \alpha_i \leq 1$ for all i. Let I be any subinterval of the unit interval. The symbol I will be used to denote either the subinterval or the length thereof, the context indicating which. Let $n(I)$ denote the number of those points $\alpha_1, \alpha_2, \alpha_3, \cdots, \alpha_n$ that lie in the interval I. The sequence S is said to be *uniformly distributed* in the unit interval if for every subinterval I,

$$\lim_{n \to \infty} \frac{n(I)}{n} = I.$$

Let R be any sequence of real numbers, β_1, β_2, β_3, $\cdots$, and recall the notation $(\beta) = \beta - [\beta]$ for the fractional part of β. The sequence R is said to be *uniformly distributed modulo* 1 if the sequence (β_1), (β_2), (β_3), $\cdots$ is uniformly distributed in the unit interval.

THEOREM 6.3.   *If ξ is irrational, then the sequence ξ, 2ξ, 3ξ, 4ξ, $\cdots$ is uniformly distributed modulo* 1.

The conclusion is clearly false if ξ is rational; so the theorem could be stated as a necessary and sufficient condition for irrationality. First we give a proof by continued fractions, and in the next section an alternative proof by Fourier series.

Proof. Let any positive ϵ and any subinterval I of the unit interval be given. As earlier, let $n(I)$ denote the number of those points

$$(\xi), (2\xi), (3\xi), \cdots, (n\xi)$$

that lie in I. We prove that

(6.2) $$\left| \frac{n(I)}{n} - I \right| < \epsilon$$

for all sufficiently large n.

Let the convergents of the continued fraction expansion of ξ be h_0/k_0, h_1/k_1, $\cdots$. By Theorem 5.11 the k_i are increasing positive integers, and so we can locate n, if not trivially small, by the inequalities

(6.3) $$k_i k_{i-1}^{\frac{1}{2}} \leqq n < k_{i+1} k_i^{\frac{1}{2}}.$$

This relation gives i as a function of n, and, as n tends to infinity, so do i and k_i. We shall prove (6.2) for all n large enough so that the corresponding i value satisfies

(6.4) $$k_{i-1}^{-\frac{1}{2}} < \frac{\epsilon}{8} \quad \text{and} \quad 2k_i^{-\frac{1}{2}} < I.$$

The idea of the proof is to work with the convergents h_i/k_i in place of ξ. This we do by using a subinterval I' of I defined as follows: I' is the interval I with a segment of length $k_i^{-1/2}$ removed from each end. The equation for lengths is thus

(6.5) $$I' = I - 2k^{-1/2}.$$

Define q by

(6.6) $q = [n/k_i]$, so that $qk_i \leqq n < (q+1)k_i$.

The fraction h_i/k_i is in lowest terms by Lemma 5.5 so that the k_i integers

$$h_i, 2h_i, \cdots, k_ih_i$$

form a complete residue system modulo k_i. Thus the points

$$(mh_i/k_i) \quad \text{for} \quad m = 1, 2, \cdots, k_i$$

are merely some arrangement of the points

(6.7) $$0, \frac{1}{k_i}, \frac{2}{k_i}, \cdots, \frac{k_i - 1}{k_i}.$$

These points being uniformly spaced, we can say that at least $k_iI' - 1$ of them lie in the interval I', the 1 being subtracted to allow for end effects. Next, the points

$$(mh_i/k_i) \quad \text{for} \quad m = 1, 2, \cdots, qk_i$$

are the points (6.7) in some order repeated q times, so that, of these qk_i points, at least $q(k_iI' - 1)$ lie in the interval I'.

We now prove that, for any value of m for which (mh_i/k_i) is a point in I', the corresponding point $(m\xi)$ lies in I. Using Lemma 5.12, (6.3), and (6.6), we see that

$$\left| m\xi - \frac{mh_i}{k_i} \right| = m \left| \xi - \frac{h_i}{k_i} \right| < \frac{qk_i}{k_ik_{i+1}} \leqq \frac{n}{k_ik_{i+1}} < k_i^{-1/2}.$$

This inequality says that the distance between the points $m\xi$ and mh_i/k_i is less than $k_i^{-\frac{1}{2}}$, and, from the definition of I', we see that $m\xi$ and mh_i/k_i do not span an integer. Hence the distance from $(m\xi)$ to (mh_i/k_i) is also less than $k_i^{-\frac{1}{2}}$, so that $(m\xi)$ belongs to I. This enables us to reformulate the conclusion of the last paragraph thus: that, of the points (ξ), (2ξ), $\cdots$, $(qk_i\xi)$, at least $q(k_iI' - 1)$ belong to the interval I. But $n \geqq qk_i$ by (6.6), and hence we have

$$n(I) \geqq q(k_iI' - 1).$$

Application of (6.5) and (6.6) yields, since $I \leqq 1$,

$$n(I) \geqq qk_iI' - q > (n - k_i)(I - 2k_i^{-\frac{1}{2}}) - nk_i^{-1}$$
$$> nI - k_i - 3nk_i^{-\frac{1}{2}},$$
$$\frac{n(I)}{n} > I - \frac{k_i}{n} - 3k_i^{-\frac{1}{2}}.$$

By (6.3) and (6.4) we have

$$\frac{k_i}{n} \leqq k_{i-1}^{-\frac{1}{2}} < \frac{\epsilon}{8} \quad \text{and} \quad 3k_i^{-\frac{1}{2}} < 3k_{i-1}^{-\frac{1}{2}} < \frac{3\epsilon}{8},$$

and these can be applied to the previous result to give

$$(6.8) \qquad \frac{n(I)}{n} - I > -\frac{\epsilon}{2}.$$

Denote by C the complement of I in the unit interval: If I is the interval (a, b), C would consist of the intervals $(0, a)$ and $(b, 1)$. The inequality (6.8), interpreted for C, would be written

$$\frac{n(C)}{n} - C > -\epsilon,$$

where $n(C)$ is the number of those points (ξ), (2ξ), $\cdots$, $(n\xi)$ that lie in the intervals C. The change from $\epsilon/2$ to

ϵ is to account for two intervals making up C. It is clear that $n(I) + n(C) = n$, and the lengths satisfy the relation $I + C = 1$. Thus we can get rid of C in the last inequality,

$$\frac{n - n(I)}{n} - 1 + I > -\epsilon \quad \text{or} \quad \frac{n(I)}{n} - I < \epsilon.$$

This, together with (6.8), implies (6.2), and the theorem is proved.

COROLLARY 6.4. *The number ξ is irrational if and only if the points (ξ), (2ξ), (3ξ), $\cdots$ are everywhere dense in the unit interval.*

Proof. If ξ is irrational, the conclusion follows at once from Theorem 6.3. If ξ is rational, say $\xi = a/b$ with $b > 0$ and $(a, b) = 1$, then the point set is discrete. It is in fact the finite set of points, 0, $1/b$, $2/b$, $\cdots$, $(b - 1)/b$.

4. A proof by Fourier analysis. We now give a more sophisticated proof of Theorem 6.3 by the use of trigonometric sums. The method, due to H. Weyl, has been used extensively in the theory of numbers since it was first introduced. This section is in effect an appendix to the chapter since we assume a requisite proposition concerning Fourier series. Thus we state the following result without proof; it may be found, for example, in Dunham Jackson, *Fourier Series and Orthogonal Polynomials*, Carus Monograph 6, p. 21.

LEMMA 6.5. *The Fourier series for a broken-line function converges uniformly to the function for all values of the variable.*

Note: By a "*broken-line function*" is meant a continuous periodic function whose graph in any one period interval consists of a finite number of straight-line segments of finite slope.

The next result is a special form of Weyl's criterion for uniform distribution.

THEOREM 6.6.   *The sequence of real numbers* $\beta_1, \beta_2, \cdots$ *is uniformly distributed modulo* 1 *if*

$$(6.9) \qquad \lim_{n \to \infty} \frac{1}{n} \sum_{j=1}^{n} \exp\left(2\pi i h \beta_j\right) = 0$$

for every positive integer h.

Proof.   Let γ be any real number satisfying $0 < \gamma \leqq 1$, and let I denote the interval of points t such that $0 \leqq t < \gamma$.   Recalling the notation (β) for the fractional part of β, we let $n(I)$ denote the number of points among (β_1), (β_2), $\cdots$, (β_n) that lie in I.   Assuming (6.9), we prove that

$$(6.10) \qquad \lim_{n \to \infty} \frac{n(I)}{n} = \gamma.$$

This result, although special in that it relates only to the special intervals having zero as left end point, clearly implies the general result for any subinterval of the unit interval, inasmuch as any subinterval is expressible as the difference of two of these special intervals with left end point zero.   The result (6.10) is obvious in case $\gamma = 1$; so we may assume that $\gamma < 1$.

Define the periodic function $g(t)$ by

$$g(t) = 1 \quad \text{for} \quad 0 \leqq t < \gamma; \qquad g(t) = 0 \quad \text{for} \quad \gamma \leqq t < 1;$$

$$g(t + k) = g(t)$$

for any integer k.   Then we have

$$(6.11) \qquad n(I) = \sum_{j=1}^{n} g(\beta_j).$$

Now $g(t)$ is not continuous. In order to get two continuous functions $g_1(t)$ and $g_2(t)$ which approximate $g(t)$ both below and above, we proceed as follows. Let ϵ be any positive real number, sufficiently small so that $2\epsilon < \gamma$ and $2\epsilon < 1 - \gamma$. Define the periodic, continuous, broken-line functions $g_1(t)$ and $g_2(t)$ (see the accompanying figure) by the equations:

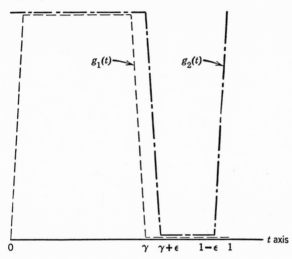

$g_1(t + k) = g_1(t)$ and $g_2(t + k) = g_2(t)$ for any integer k;

$g_1(t) = 1$ for $\epsilon \leqq t \leqq \gamma - \epsilon$;

$g_1(t) = 0$ for $\gamma \leqq t \leqq 1$;

$g_1(t)$ linear for $0 \leqq t \leqq \epsilon$ and $\gamma - \epsilon \leqq t \leqq \gamma$;

$g_2(t) = 1$ for $0 \leqq t \leqq \gamma$;

$g_2(t) = 0$ for $\gamma + \epsilon \leqq t \leqq 1 - \epsilon$;

$g_2(t)$ linear for $\gamma \leqq t \leqq \gamma + \epsilon$ and $1 - \epsilon \leqq t \leqq 1$.

Note that for all values of t

$$(6.12) \qquad g_1(t) \leqq g(t) \leqq g_2(t).$$

By Lemma 6.5 both $g_1(t)$ and $g_2(t)$ have uniformly convergent Fourier series expansions; say

$$g_1(t) = a_0 + \sum_{h=1}^{\infty} (a_h \cos 2\pi ht + b_h \sin 2\pi ht),$$

(6.13)

$$g_2(t) = c_0 + \sum_{h=1}^{\infty} (c_h \cos 2\pi ht + d_h \sin 2\pi ht).$$

We shall need the actual values of a_0 and c_0; thus

(6.14)

$$a_0 = \int_0^1 g_1(t)\, dt = \gamma - \epsilon, \qquad c_0 = \int_0^1 g_2(t)\, dt = \gamma + \epsilon.$$

As for the other coefficients, it will suffice to have the rough inequalities

$$(6.15) \quad |a_h| \leqq 2, \quad |b_h| \leqq 2, \quad |c_h| \leqq 2, \quad |d_h| \leqq 2,$$

for all integers $h \geqq 1$; these can be proved very simply by direct estimates: for example

$$|a_h| = \left| 2 \int_0^1 g_1(t) \cos 2\pi ht\, dt \right|$$

$$\leqq 2 \max |g_1(t)| \cdot |\cos 2\pi ht| \leqq 2.$$

Now the uniform convergence of the series in (6.13) implies the existence of an integer M such that for any value of t

(6.16)
$$\left| \sum_{h=m+1}^{\infty} (a_h \cos 2\pi ht + b_h \sin 2\pi ht) \right| < \epsilon$$

for any $m \geqq M$.

Turning to (6.9), we rewrite it in the form

$$\lim_{n \to \infty} \frac{1}{n} \sum_{j=1}^{n} (\cos 2\pi h\beta_j + i \sin 2\pi h\beta_j) = 0.$$

The limit of both the real and imaginary parts must be zero; so there exists an integer N so that

(6.17)

$$\left| \frac{1}{n} \sum_{j=1}^{n} \cos 2\pi h\beta_j \right| < \frac{\epsilon}{M} \quad \text{and} \quad \left| \frac{1}{n} \sum_{j=1}^{n} \sin 2\pi h\beta_j \right| < \frac{\epsilon}{M}$$

for all $h = 1, 2, \cdots, M$ and for all $n \geqq N$. Now let n be any integer satisfying both $n \geqq N$ and $n \geqq M$. Replace t successively by $\beta_1, \beta_2, \cdots, \beta_n$ in $g_1(t)$ in (6.13), and add the resulting equations to get, by (6.14),

$$\sum_{=1}^{n} g_1(\beta_j) = n(\gamma - \epsilon)$$
$$+ \sum_{j=1}^{n} \sum_{h=M+1}^{\infty} (a_h \cos 2\pi h\beta_j + b_h \sin 2\pi h\beta_j)$$
$$+ \sum_{j=1}^{n} \sum_{h=1}^{M} (a_h \cos 2\pi h\beta_j + b_h \sin 2\pi h\beta_j).$$

We reverse the order of summation in the finite sums in the last term. Then we apply the triangle inequality and (6.16) and (6.17). Since both $g_1(t)$ and $\gamma - \epsilon$ are non-negative, we get

$$\sum_{j=1}^{n} g_1(\beta_j)$$

$$= \left| \sum_{j=1}^{n} g_1(\beta_j) \right|$$

$$\geq n(\gamma - \epsilon) - \sum_{j=1}^{n} \left| \sum_{h=M+1}^{\infty} (a_h \cos 2\pi h \beta_j + b_h \sin 2\pi h \beta_j) \right|$$

$$- \sum_{h=1}^{M} \left| a_h \sum_{j=1}^{n} \cos 2\pi h \beta_j + b_h \sum_{j=1}^{n} \sin 2\pi h \beta_j \right|$$

$$\geq n(\gamma - \epsilon) - \sum_{j=1}^{n} \epsilon - \sum_{h=1}^{M} \left\{ |a_h| \cdot \left| \sum_{j=1}^{n} \cos 2\pi h \beta_j \right| \right.$$

$$\left. + |b_h| \cdot \left| \sum_{j=1}^{n} \sin 2\pi h \beta_j \right| \right\}$$

$$\geq n(\gamma - \epsilon) - n\epsilon - \sum_{h=1}^{M} \left\{ 2 \cdot \frac{n\epsilon}{M} + 2 \cdot \frac{n\epsilon}{M} \right\}$$

$$= n(\gamma - 6\epsilon).$$

A similar analysis of $g_2(t)$ leads to the conclusion that

$$\sum_{j=1}^{n} g_2(\beta_j) \leq n(\gamma + 6\epsilon)$$

for all sufficiently large n. Next we use (6.12) in conjunction with these inequalities to obtain

$$\gamma - 6\epsilon \leq \frac{1}{n} \sum_{j=1}^{n} g(\beta_j) \leq \gamma + 6\epsilon.$$

But ϵ is arbitrarily small, and hence by (6.11)

$$\lim_{n \to \infty} \frac{n(I)}{n} = \lim_{n \to \infty} \frac{1}{n} \sum_{j=1}^{n} g(\beta_j) = \gamma,$$

which is (6.10).

Having completed the proof of Theorem 6.6, we now deduce Theorem 6.3 from it. We must establish that (6.9) holds for the values $\beta_j = j\xi$, where ξ is any irrational number. We note that

$$\left| \sum_{j=1}^{n} \exp (2\pi i h j \xi) \right|$$

$$= \frac{|\exp (2\pi i h \xi(n+1)) - \exp (2\pi i h \xi)|}{|\exp (2\pi i h \xi) - 1|}$$

$$\leq \frac{|\exp (2\pi i h \xi(n+1))| + |\exp (2\pi i h \xi)|}{|\exp (2\pi i h \xi) - 1|}$$

$$\leq \frac{2}{|\exp (2\pi i h \xi) - 1|} = c(h, \xi),$$

where $c(h, \xi)$ is obviously independent of n; also note that $|\exp (2\pi i h \xi) - 1| \neq 0$ since $h\xi$ is irrational. Hence we obtain

$$\lim_{n \to \infty} \frac{1}{n} \left| \sum_{j=1}^{n} \exp (2\pi i h j \xi) \right| \leq \lim_{n \to \infty} \frac{c(h, \xi)}{n} = 0,$$

which implies (6.9) with β_j replaced by $j\xi$.

Notes on Chapter 6

Theorems 6.1 and 6.2 are due to A. Hurwitz, *Math. Annalen*, **39** (1891), 279–284. These results have been extended to the complex case as follows by L. R. Ford, *Trans. Amer. Math. Soc.*, **27** (1925), 146–154. We say that $\alpha = a + bi$ is a complex irrational number if not both a and b are rational, and that $x = x_1 + x_2 i$ is a complex integer, or Gaussian integer, in case both x_1 and x_2 are rational integers. If α is any complex irrational number, then there exist infinitely many pairs x, y of complex integers such that

$$\left| \alpha - \frac{y}{x} \right| < \frac{1}{\sqrt{3} \, |x|^2}.$$

This result is best possible in the sense that the constant $\sqrt{3}$ cannot be replaced by any larger value; for example, $\alpha = (-1 + i\sqrt{3})/2$ is a critical number requiring the $\sqrt{3}$ in the above approximation, in the same way that $(\sqrt{5} + 1)/2$ served as a critical number in the proof of Theorem 6.2. Ford's proofs are along geometric lines quite different from the Hurwitz method of continued fractions, which does not seem to generalize directly to the complex case.

Theorem 6.3 was proved independently at about the same time by P. Bohl, W. Sierpinski, and H. Weyl; for details see J. F. Koksma, *Diophantische Approximationen*, p. 92. Theorem 6.3 implies the following result of Kronecker: given any irrational θ, any real α, and any positive ϵ, there exist infinitely many pairs of integers n, m such that

$$|n\theta - m - \alpha| < \epsilon.$$

A more general form of Kronecker's theorem is given in Hardy and Wright, Chapter XXIII. Generalizations of both Kronecker's theorem and Theorem 6.3 may be found in papers of H. Weyl, *Math. Annalen*, **77** (1916), 313–352, and J. G. van der Corput, *Acta Math.*, **56** (1931), 373–456.

The work of Weyl just cited is the source of Theorem 6.6. The converse of this theorem also holds; that is, the condition (6.9) is necessary as well as sufficient for the uniform distribution of the β's. This converse proposition, and the fairly easy proof thereof, were not included in the chapter because they are not relevant to our purposes.

CHAPTER 7

ALGEBRAIC AND
TRANSCENDENTAL NUMBERS

1. Closure properties of algebraic numbers. Algebraic and transcendental numbers were defined at the beginning of Chapter 3, and a few basic ideas were outlined. It is customary to separate the *complex* numbers into the types algebraic and transcendental, whereas it is the *real* numbers that are classified as rational and irrational. These usages are readily extended, of course. We would say that $a + bi$ is a complex rational if both a and b are rational. Likewise a real algebraic number is simply one that is simultaneously real and algebraic. If a complex number is algebraic, must its real and complex parts be algebraic? The answer is yes, and to establish this we first prove the basic closure properties of algebraic numbers.

LEMMA 7.1. *Any $r + 1$ linear forms in r indeterminates with rational coefficients are linearly dependent over the rationals.*

Remark. This is a well-known result, and so we give a reference but not a proof. The matrix of the linear forms cannot have rank greater than r, and this is less than the number of forms. For further details, see any book on the theory of equations, for example, J. M. Thomas, *Theory of Equations*, Theorem 27.7, p. 49, McGraw-Hill.

THEOREM 7.2. *Algebraic numbers are closed under addition and multiplication; in other words, the sum and the product of two algebraic numbers are themselves algebraic. More generally, algebraic numbers form a field.*

Proof. Let α and β be algebraic numbers of degrees m and n, respectively. Thus α satisfies an algebraic equation of degree m, say

$$(7.1) \quad \alpha^m = a_{m-1}\alpha^{m-1} + a_{m-2}\alpha^{m-2} + \cdots + a_1\alpha + a_0,$$

with rational coefficients a_j. Thus α^m is a linear combination of $1, \alpha, \alpha^2, \cdots, \alpha^{m-1}$ with rational coefficients. The same is true for α^{m+1}, as can be seen by multiplying equation (7.1) by α and then using (7.1) to replace the term $a_{m-1}\alpha^m$ by terms in α of lower degree. This process can be repeated, and so by mathematical induction we see that each of $\alpha^m, \alpha^{m+1}, \alpha^{m+2}, \cdots$ is expressible as a linear combination of $1, \alpha, \cdots, \alpha^{m-1}$ with rational coefficients. Similarly, each of $\beta^n, \beta^{n+1}, \beta^{n+2}, \cdots$ is expressible as a linear combination of $1, \beta, \beta^2, \cdots, \beta^{n-1}$ with rational coefficients.

Now consider the $mn + 1$ numbers

$$(7.2) \qquad 1, \alpha + \beta, (\alpha + \beta)^2, \cdots, (\alpha + \beta)^{mn}.$$

Expanding these and replacing the mth and higher powers of α by the lower powers, and likewise the nth and higher powers of β, we see that these $mn + 1$ numbers can be written as linear combinations of the mn numbers

$$\alpha^j\beta^k, \qquad j = 0, 1, \cdots, m - 1, \qquad k = 0, 1, \cdots, n - 1,$$

with rational coefficients. These mn numbers can be substituted for the r indeterminates of Lemma 7.1, and so we conclude that the numbers (7.2) are linearly independent over the rationals. That is, $\alpha + \beta$ is an algebraic number.

The proof for the product $\alpha\beta$ is analogous, the powers of $\alpha + \beta$ in (7.2) being replaced by the powers of $\alpha\beta$.

It is clear that algebraic numbers are also closed under subtraction and division, except division by zero. For, if α satisfies $f(x) = 0$, a polynomial equation of degree m, then $-\alpha$ and α^{-1} satisfy $f(-x) = 0$ and $x^m f(1/x) = 0$, both polynomial equations.

COROLLARY 7.3. *For real values a and b, the number $a + bi$ is algebraic if and only if a and b are algebraic.*

Proof. The number i is algebraic since it is a root of $x^2 + 1 = 0$. Hence, if a and b are algebraic, so is $a + bi$ by Theorem 7.2. Conversely, if $a + bi$ is algebraic, say $f(a + bi) = 0$ where $f(x)$ is a polynomial with rational coefficients, then $a - bi$ is also a root of $f(x) = 0$. Hence the sum $2a$ and the difference $2bi$ of $a + bi$ and $a - bi$ are also algebraic numbers, and so, multiplying by the algebraic numbers $1/2$ and $-i/2$, we see that a and b are algebraic.

2. A property of algebraic integers. It was established in § 3 of Chapter 1 that the rational numbers are everywhere dense. Since the rational numbers form a subset of the real algebraic numbers, it follows that these also are everywhere dense. Indeed we can say much more.

THEOREM 7.4. *The real algebraic integers of any specified degree $n \geq 2$ are everywhere dense on the real line.*

Proof. Let α and β be any two distinct real numbers: say with $\alpha < \beta$. We must prove that there is a real algebraic integer γ of degree n such that $\alpha < \gamma < \beta$. First note that

$$(x + \beta)^n - (x + \alpha)^n = \{(x + \alpha) + (\beta - \alpha)\}^n$$
$$- (x + \alpha)^n > n(x + \alpha)^{n-1}(\beta - \alpha)$$

for all x such that $x + \alpha > 0$. The above inequality was obtained by discarding terms past the second in the binomial expansion.   Now the value of $n(x + \alpha)^{n-1}(\beta - \alpha)$ can be made arbitrarily large by choosing x large.   Hence there is a positive integer j such that

$$(j + \beta)^n - (j + \alpha)^n > 5, \qquad j + \alpha > 0, \qquad j + \beta > 0.$$

Thus the open interval from $(j + \alpha)^n$ to $(j + \beta)^n$ contains at least four consecutive positive integers, and so in particular a positive integer of the form $2 + 4k$.   By the continuity principle there is a real number γ such that $(j + \gamma)^n = 2 + 4k$, with $\alpha < \gamma < \beta$.   It follows that the real number γ is an algebraic integer, since it is a root of the monic polynomial equation

$$f(x) = (x + j)^n - 2(1 + 2k) = 0,$$

whose coefficients are rational integers.

To complete the proof we must establish that the degree of the algebraic integer γ is n: i.e., that the polynomial $f(x)$ is irreducible over the rationals.   This is equivalent to proving the irreducibility of $f(x - j) = x^n - 2(1 + 2k)$. The simplest way of handling this would be to cite the Eisenstein irreducibility criterion (cf. glossary), but the following direct approach is easy enough.   The zeros of the polynomial $f(x - j)$ are the complex numbers

$$\sqrt[n]{2(1 + 2k)} \cdot \zeta^s, \qquad s = 1, 2, \cdots, n,$$

where ζ is a primitive nth root of unity.   Assuming that $f(x - j)$ were reducible, it would have a polynomial factor $g(x)$ with rational coefficients, say of positive degree $w < n$.   Then $g(x)$ would be the product of w linear factors of the form

$$x - \sqrt[n]{2(1 + 2k)} \cdot \zeta^s.$$

Consider the absolute value of the constant term of such a polynomial.   It would be $\{2(1 + 2k)\}^{w/n}$, since the absolute value of ζ^s is 1 for any value of s. But under our assumption this would be rational, say a/b,

$$\{2(1 + 2k)\}^{w/n} = a/b \quad \text{or} \quad 2^w(1 + 2k)^w b^n = a^n.$$

The latter equation is impossible, because the power of 2 dividing the integer on the right cannot equal the power of 2 dividing the left side.

3. Transcendental numbers. In § 2 of Chapter 1 it was proved that the rational numbers are countable. This is now extended to the algebraic numbers.

THEOREM 7.5. *The set of all algebraic numbers is countable; likewise the set of all real algebraic numbers.*

Proof. For our purposes here we think of any algebraic number as satisfying some polynomial equation

$$f(x) = a_0x^n + a_1x^{n-1} + \cdots + a_n = 0$$

with rational integral coefficients, $a_0 \neq 0$. There is no loss of generality in taking $a_0 \geq 1$. Define the index of the polynomial $f(x)$ to be the positive integer

$$n + a_0 + |a_1| + |a_2| + \cdots + |a_n|.$$

Since $n \geq 1$ and $a_0 \geq 1$, the index is at least 2 for any polynomial. In fact there is only one polynomial of index 2: namely x. This gives the algebraic number 0 as a root of the equation $f(x) = x = 0$. The polynomials of index 3 are x^2, $x + 1$, $x - 1$, $2x$. These give the new algebraic numbers ± 1. Similarly the polynomials of index 4 give new algebraic numbers ± 2, $\pm\frac{1}{2}$, $\pm i$. Clearly there is only a finite number of polynomials of any given index, and so only a finite number of algebraic numbers given thereby. By allowing the index to run through the natural numbers, and listing the new algebraic numbers obtained at each stage, we would have enumerated a sequence of distinct algebraic numbers. Furthermore, all algebraic numbers would be in the sequence, because every polynomial has an index. Thus the algebraic numbers are countable. If at each stage we had listed only the *real* algebraic numbers, we would obtain the second conclusion stated in the theorem.

COROLLARY 7.6. *Almost all real numbers are transcendental.*

Proof. By Theorem 7.5 the real algebraic numbers are countable, and so have measure zero by Theorem 1.3. The corollary follows at once, by the definition of "almost all."

This corollary can be readily generalized to the complex case: almost all complex numbers are transcendental. To give meaning to this, we must extend the definitions of "measure zero" and "almost all" to the complex case. A set in the complex plane is said to have *measure zero* if it can be covered with neighborhoods (circles, for example) of arbitrarily small total area. Thus any countable set has measure zero, by an argument analogous to that in the proof of Theorem 1.3. *Almost all* complex numbers are said to have property P in case the set of complex numbers (or points in the complex plane) which lack the property P has measure zero.

4. The order of approximation. A real number ξ is said to be *approximable* by rationals *to order n* if there exists a positive constant c depending only on ξ such that the inequality

$$(7.3) \qquad\qquad \left| \xi - \frac{h}{k} \right| < \frac{c}{k^n}$$

has infinitely many rational solutions h/k with $k > 0$ and $(h, k) = 1$.

If a number is approximable say to order 5, then it is a fortiori approximable to any smaller order, 4, 3, 2, or 1. Theorem 4.1 implies that any irrational is approximable to order 2; in this case the constant c of (7.3) would be 1. The same conclusion can be drawn from the stronger Theorem 6.1; this theorem would enable us to improve the constant c to $1/\sqrt{5}$, but from our present viewpoint this improvement is immaterial. This result for irrational numbers should be contrasted with the following proposition on rationals.

THEOREM 7.7. *Any rational number is approximable to order* 1, *but not to any higher order.*

Proof. Consider any rational number a/b with $(a, b) = 1$, $b \geqq 1$. It is well known from elementary number theory that there are infinitely many integral solutions of the equation $ax - by = 1$. In fact, if $x = x_0$, $y = y_0$ is one such solution, then the general solution is $x = x_0 + bt$, $y = y_0 + at$, where t ranges over all integers. Whatever the values of x_0 and b, it is clear that $x_0 + bt$ is positive for infinitely many values of the integer t. Thus there are infinitely many integral solutions x, y with $x > 0$ to the formulas

$$ax - by = 1, \qquad \left| \frac{a}{b} - \frac{y}{x} \right| = \frac{1}{bx}, \qquad \left| \frac{a}{b} - \frac{y}{x} \right| < \frac{2}{x}.$$

This inequality can be interpreted as saying that the rational a/b is approximable to order 1, for it is inequality (7.3) with $n = 1$, $\xi = a/b$, $h/k = y/x$, and $c = 2$.

On the other hand, for any fraction $y/x \neq a/b$ we see that

$$\left| \frac{a}{b} - \frac{y}{x} \right| = \left| \frac{ax - by}{bx} \right| \geqq \frac{1}{bx}.$$

There is no constant c such that $1/bx < c/x^2$ for infinitely many integers x, and so a/b is not approximable to order higher than 1.

Theorem 7.7 can be used to prove the irrationality of such a number as

$$\xi = 10^{-1} + 10^{-2} + 10^{-4} + 10^{-8} + \cdots + 10^{-2^m} + \cdots,$$

in the following way. Define the rational numbers r_m by means of

$$r_m = 10^{-1} + 10^{-2} + 10^{-4} + 10^{-8} + \cdots + 10^{-2^m},$$

so that r_m can be thought of as a fraction with denominator 10^{2^m}. Then we can write

$$|\xi - r_m| = 10^{-2^{m+1}} + 10^{-2^{m+2}} + \cdots < 2 \cdot 10^{-2^{m+1}}$$

$$= 2(10^{-2^m})^2.$$

This can be interpreted as an inequality of the form (7.3), with $h/k = r_m$, $n = 2$, $k = 10^{2^m}$, $c = 2$. Since there are infinitely many r_m as m runs through the natural numbers, ξ is approximable to order 2, and hence is irrational by Theorem 7.7. The argument just completed will be generalized after Theorem 7.8, following the procedure of Liouville, to exhibit a class of transcendental numbers.

THEOREM 7.8. *A real algebraic number of degree n is not approximable to order $n + 1$ or higher.*

Proof. Theorem 7.7 covers the case $n = 1$, and so we take $n \geq 2$. Let the algebraic number ξ of degree n satisfy the equation, irreducible over the rationals,

$$f(x) = a_0 x^n + a_1 x^{n-1} + \cdots + a_n = 0,$$

with integral coefficients. For any x in the range $\xi - 1 < x < \xi + 1$ we have $|x| < |\xi| + 1$, and hence the derivative $f'(x)$ satisfies

(7.4)

$$|f'(x)| = |na_0x^{n-1} + (n-1)a_1x^{n-2} + \cdots + a_{n-1}|$$
$$\leqq |na_0x^{n-1}| + |(n-1)a_1x^{n-2}| + \cdots + |a_{n-1}|$$
$$< n|a_0|\{|\xi|+1\}^{n-1} + (n-1)|a_1|\{|\xi|+1\}^{n-2}$$
$$+ \cdots + |a_{n-1}| = A.$$

This equality defines the constant A. Now, if h/k is any rational approximation to ξ, we may presume that $\xi - 1 < h/k < \xi + 1$. Also $f(h/k) \neq 0$ since otherwise $f(x)$ would have the factor $x - h/k$, contrary to the irreducibility of $f(x)$ over the rationals. It follows that

$$\left| f\left(\frac{h}{k}\right) \right| = \frac{|a_0h^n + a_1h^{n-1}k + \cdots + a_nk^n|}{k^n} \geqq \frac{1}{k^n}.$$

By the law of the mean we have

$$f\left(\frac{h}{k}\right) = f\left(\frac{h}{k}\right) - f(\xi) = \left(\frac{h}{k} - \xi\right)f'(x)$$

for some x between h/k and ξ. Taking absolute values in this equation, we apply the last inequality and (7.4) to get

$$\left| \frac{h}{k} - \xi \right| = \frac{|f(h/k)|}{|f'(x)|} > \frac{1}{Ak^n}.$$

There is no constant c so that $1/(Ak^n) < c/k^{n+1}$ for infinitely many positive integers k, and consequently ξ is not approximable to order $n + 1$ or higher.

DEFINITION. A real number ξ is said to be a *Liouville number* if for every positive integer m there is a rational number h_m/k_m with $k_m > 1$ such that

(7.5) $$|\xi - h_m/k_m| < (k_m)^{-m}.$$

THEOREM 7.9. *Any Liouville number is transcendental.*

Proof. Assume that a Liouville number ξ is algebraic of degree n. For all integers $m \geqq n + 1$ the inequality (7.5) would imply that

$$|\xi - h_m/k_m| < (k_m)^{-n-1},$$

and thus ξ would be approximable to order $n + 1$. But this contradicts Theorem 7.8.

Theorem 7.9 can be used to construct transcendental numbers. For example, consider

$$\xi_1 = 10^{-1!} + 10^{-2!} + \cdots + 10^{-m!} + \cdots = 0.1100010 \cdots,$$

and define h_m/k_m as the sum of the first m terms of the infinite series, so that $k_m = 10^{m!}$. Then we have

$$
\begin{aligned}
|\xi_1 - h_m/k_m| &= 10^{-(m+1)!} + 10^{-(m+2)!} + \cdots \\
&< 2 \cdot 10^{-(m+1)!} < (10^{-m!})^{-m} \\
&= (k_m)^{-m}.
\end{aligned}
$$

Thus the conditions (7.5) are satisfied, and ξ_1 is transcendental.

This single number ξ_1 can be readily extended into a whole class of transcendental numbers. For example, we could employ the series

$$a_1 \, 10^{-1!} + a_2 \, 10^{-2!} + a_3 \, 10^{-3!} + \cdots,$$

where each a_i is either 1 or 2.

Notes on Chapter 7

Theorem 7.2 implies that the *real* algebraic numbers form a field also. Algebraic integers are closed under addition and multiplication, but this cannot be concluded from the proof of Theorem 7.2 as given; see Pollard, pp. 58–60.

The material in § 3 is essentially the work of G. Cantor, that in § 4 the work of J. Liouville. Liouville's work was the earlier, in *J. Math. Pures Appl.* (1), **16** (1851), 133–142; this paper gives the first occurrence of transcendental numbers in mathematics. Lucid introductions to the Cantor theory and the theory of sets generally are given in E. Kamke, *Theory of Sets*, and R. L. Wilder, *Introduction to the Foundations of Mathematics*.

Much more is known about the approximation of algebraic numbers than is indicated in Theorem 7.8. The following theorem of K. F. Roth, *Mathematika*, **2** (1955), part 1, no. 3, 1–20, is a refinement of earlier results by F. J. Dyson, C. L. Siegel, and A. Thue. If ξ is an algebraic number, not rational, and if there are infinitely many rational numbers h/k such that

$$\left| \xi - \frac{h}{k} \right| < k^{-\mu},$$

then $\mu \leq 2$. This inequality is best possible, in view of Theorem 4.1.

There is an account of the Thue–Siegel result mentioned above in L. E. Dickson, *Introduction to the Theory of Numbers*, University of Chicago Press, 1929, Theorems 108 and 111. This account also gives (Theorem 107) the following result of Thue which is implied by his approximation theorem: for any integer c the Diophantine equation $H(x, y) = c$ has at most a finite number of solutions in integers if $H(x, y)$ is a homogeneous polynomial of degree $n \geq 3$,

$$H(x, y) = a_0 x^n + a_1 x^{n-1} y + a_2 x^{n-2} y^2 + \cdots + a_n y^n,$$

with integral coefficients such that $H(x, 1)$ is irreducible over the field of rational numbers.

CHAPTER 8

NORMAL NUMBERS

1. Definition of a normal number. To put the matter roughly at first, a normal number is one in whose decimal expansion all digits occur with equal frequency, and in fact all blocks of digits of the same length occur with equal frequency. For example, if the base is 10, then such a specified digit as 7 occurs with frequency 10^{-1}, and any triplet of digits such as 357 occurs with frequency 10^{-3}. As we shall see, almost all numbers have this property, which somewhat justifies the use of the adjective "normal."

We now formulate the definition exactly. Let $x = .x_1 x_2 x_3 \cdots$ be an infinite decimal to base r, and let X_n denote the block of digits $x_1 x_2 \cdots x_n$. Thus each digit x_i has one of the values $0, 1, \cdots, r - 1$. For any particular value b among these r possibilities, let $N(b, X_n)$ denote the number of occurrences of b in the block X_n. For example, if $X_6 = 123113$, then $N(1, X_6) = 3$, $N(2, X_6) = 1$, $N(3, X_6) = 2$, $N(0, X_6) = 0$, etc. To illustrate further the meaning of this notation, we mention that

$$\sum_{b=0}^{r-1} N(b, X_n) = n.$$

The number x is simply normal to base r if

$$(8.1) \qquad \lim_{n \to \infty} \frac{1}{n} N(b, X_n) = \frac{1}{r}$$

for each of the r values of b. Thus x is simply normal to base r if each digit b occurs with frequency $1/r$.

The number x is *normal to base r* if each block B_k of k digits occurs with frequency $1/r^k$. To spell this out in more detail, we let $N(B_k, X_n)$ denote the number of occurrences of the block $B_k = b_1 b_2 \cdots b_k$ in X_n. Then x is normal to base r if

$$(8.2) \qquad \lim_{n \to \infty} \frac{1}{n} N(B_k, X_n) = r^{-k}$$
$$\text{for all} \quad B_k, \qquad k = 1, 2, 3, \cdots.$$

Finally, if y is any real number, y is said to be normal to base r if $y - [y]$ is normal to base r.

These definitions are in no way invalidated by the ambiguous decimal representation of certain rational numbers. For example, consider the number $y = 0.03000\cdots = 0.02999\cdots$ to base 10. This particular number y is not simply normal to base 10, and so not normal to base 10, regardless of which of its two decimal expansions is examined. Likewise, any number that has two decimal expansions to base r is not simply normal to that base.

An example of a number that is simply normal but not normal to base 10 is given by

$$0.0123456789012345678901234567 89 \cdots.$$

In fact, since any rational number has a periodic decimal expansion whatever the base, it follows that normal numbers are irrational.

The above definition of a normal number is equivalent to the original definition of E. Borel, according to whom x is normal to base r if each of x, rx, $r^2 x$, $\cdots$ is simply normal to all of the bases r, r^2, r^3, $\cdots$. To formulate this Borel definition as a set of equations, let us break $N(B_k, X_n)$ into k parts

$$(8.3) \qquad N(B_k, X_n) = \sum_{s=1}^{k} N_s(B_k, X_n),$$

where $N_s(B_k, X_n)$ denotes the number of occurrences of B_k in X_n, say $b_1 = x_j$, $b_2 = x_{j+1}$, etc., which satisfy the condition $j \equiv s \pmod{k}$. For example, let $X_9 = 112312121$, and we have $N(12, X_9) = 3$, $N_1(12, X_9) = 2$, $N_2(12, X_9) = 1$, $N(121, X_9) = 2$, $N_1(121, X_9) = 1$, $N_2(121, X_9) = 1$, $N_3(121, X_9) = 0$.

Borel's definition is equivalent to saying that x is normal to base r if

$$(8.4) \qquad \lim_{n \to \infty} \frac{1}{n} N_s(B_k, X_n) = k^{-1} r^{-k} \quad \text{for all} \quad k, s, B_k.$$

The proof of the equivalence of (8.2) and (8.4) will be given in § 3. However, in view of (8.3) the implication is clear in one direction, and we now state this formally for reference.

LEMMA 8.1. *If x satisfies (8.4), then x satisfies (8.2).*

S. S. Pillai established that the conditions (8.4) are redundant in part, and proved that x is normal to base r if x is simply normal to bases r, r^2, r^3, $\cdots$. That is, x is normal to base r if

$$(8.5) \qquad \lim_{n \to \infty} \frac{1}{n} N_1(B_k, X_n) = k^{-1} r^{-k}$$
$$\text{for all} \quad B_k, \qquad k = 1, 2, 3, \cdots.$$

That this is equivalent to our definition (8.2) will be established in § 3.

It will be convenient to establish some simple lemmas for application later in the chapter.

LEMMA 8.2. *Let $f_1, f_2, \cdots, f_m$ be real functions of the real variable x satisfying the conditions*

$$\lim_{x \to \infty} \{f_1(x) + f_2(x) + \cdots + f_m(x)\} = 1,$$

and

$$\liminf_{x \to \infty} f_i(x) \geqq \frac{1}{m} \quad for \quad i = 1, 2, \cdots, m.$$

Then

$$\lim_{x \to \infty} f_i(x) = \frac{1}{m} \quad for \quad i = 1, 2, \cdots, m.$$

Proof. We prove the result for the case $i = 1$, and this involves no loss of generality because of the complete symmetry in the m functions. By standard results from function theory (for example cf. L. M. Graves, *Theory of Functions of Real Variables*, p. 62, Theorem 14), we see that

$$1 = \limsup \sum_{i=1}^{m} f_i(x) \geqq \limsup f_1(x) + \liminf \sum_{i=2}^{m} f_i(x)$$

$$\geqq \limsup f_1(x) + \sum_{i=2}^{m} \liminf f_i(x)$$

$$\geqq \limsup f_1(x) + \frac{m-1}{m}.$$

This we have $\limsup f_1(x) \leqq 1/m \leqq \liminf f_1(x)$, and it follows that $\lim f_1(x) = 1/m$.

LEMMA 8.3. *The union of a countable collection of sets of measure zero is a set of measure zero.*

Proof. Let the countable collection of sets of measure zero be denoted by $S_1, S_2, S_3, \cdots$. For any given positive ϵ, the points of S_1 can be covered by a collection of intervals of length $\epsilon/2$, the points of S_2 by a collection of intervals of length $\epsilon/4$, and in general the points of S_n by a collection of intervals of length $\epsilon/2^n$. Thus the union of the sets S_n can be covered by intervals of total length not greater than $\Sigma\epsilon/2^n = \epsilon$.

2. The measure of the set of normal numbers. We now establish that almost all real numbers are normal to every base.

In the previous section we used the notation X_n to designate a fixed block of n digits. It is now convenient to regard X_n as denoting any of the r^n blocks of n digits to base r. For a fixed digit b define $p(n, j)$ as the number of blocks of digits $X_n = x_1 x_2 \cdots x_n$ having exactly j occurrences of the digit b, that is, having $N(b, X_n) = j$. Taking $0 \leqq j \leqq n$, we see that

$$(8.6) \qquad p(n, j) = \frac{n!}{j!(n-j)!}(r-1)^{n-j}.$$

It is convenient also to define $p(n, j) = 0$ in case $j < 0$ or $j > n$.

LEMMA 8.4. *If $0 < \alpha < 1$, then $1 - \alpha < \exp(-\alpha)$.*

Proof. We use the series expansion of the exponential function to write

$$\exp(-\alpha) = 1 - \alpha + \frac{\alpha^2}{2!} - \frac{\alpha^3}{3!} + \cdots > 1 - \alpha.$$

LEMMA 8.5. *For $j \geqq 2$ we have*

$$p(nr, n + j) < r^{nr} \exp\{-j^2/4nr\}.$$

Proof. From (8.6) and Lemma 8.4 we see that

$$\frac{p(nr, n+j)}{p(nr, n)}$$

$$= \frac{(nr - n)(nr - n - 1) \cdots (nr - n - j + 1)}{(n+1)(n+2) \cdots (n+j)(r-1)^j}$$

$$= \left(1 + \frac{1}{n}\right)^{-1} \left(1 + \frac{2}{n}\right)^{-1} \cdots \left(1 + \frac{j}{n}\right)^{-1}$$

$$\times \frac{(r-1)\left(r-1-\dfrac{1}{n}\right) \cdots \left(r-1-\dfrac{j-1}{n}\right)}{(r-1)^j}$$

$$< \frac{(r-1)\left(r-1-\dfrac{1}{n}\right) \cdots \left(r-1-\dfrac{j-1}{n}\right)}{(r-1)^j}$$

$$= \left\{1 - \frac{1}{n(r-1)}\right\} \left\{1 - \frac{2}{n(r-1)}\right\} \cdots$$

$$\left\{1 - \frac{j-1}{n(r-1)}\right\}$$

$$< \exp \left\{-\frac{1}{n(r-1)} - \frac{2}{n(r-1)} - \cdots - \frac{j-1}{n(r-1)}\right\}$$

$$= \exp \left\{-j(j-1)/2n(r-1)\right\} < \exp \left\{-j^2/4nr\right\}.$$

Also we have the rather obvious relation

$$(8.7) \qquad p(nr, n) < \sum_{j=0}^{nr} p(nr, j) = r^{nr},$$

and the lemma follows from these inequalities.

LEMMA 8.6.   *For $j \geqq 2$ we have*

$$p(nr, n - j) < r^{nr} \exp \left\{-j^2/4nr\right\}.$$

Proof.   Lemma 8.4 and formula (8.6) imply that

$$\frac{p(nr, n - j)}{p(nr, n)}$$

$$= \frac{n(n - 1) \cdots (n - j + 1)(r - 1)^j}{(nr - n + 1)(nr - n + 2) \cdots (nr - n + j)}$$

$$= \left(1 - \frac{1}{n}\right)\left(1 - \frac{2}{n}\right) \cdots \left(1 - \frac{j - 1}{n}\right)$$

$$\times \frac{(r - 1)^j}{(r - 1 + 1/n)(r - 1 + 2/n) \cdots (r - 1 + j/n)}$$

$$< \left(1 - \frac{1}{n}\right)\left(1 - \frac{2}{n}\right) \cdots \left(1 - \frac{j - 1}{n}\right)$$

$$< \exp\left\{-\frac{1}{n} - \frac{2}{n} - \cdots - \frac{j - 1}{n}\right\}$$

$$= \exp\left\{-j(j - 1)/2n\right\} < \exp\left\{-j^2/4nr\right\}.$$

The lemma follows from this and (8.7).

LEMMA 8.7. *Let $\epsilon > 0$ be given. For all sufficiently large n the number of blocks X_{nr} such that*

$$(8.8) \qquad |N(b, X_{nr}) - n| > n\epsilon$$

is less than $(nr)r^{nr}(1 + c_1)^{-n}$, where c_1 is a positive constant depending on ϵ but not on n.

Proof. The number of blocks X_{nr} of nr digits satisfying (8.8) is simply

$$\sum_{k > n + n\epsilon} p(nr, k) + \sum_{k < n - n\epsilon} p(nr, k) = \sum_{|j| > n\epsilon} p(nr, n + j).$$

We estimate this sum as follows. Taking n sufficiently large so that $n\epsilon > 1$, we note that the sum has fewer than nr non-zero terms. Also by the two preceding lemmas we observe that each term of the sum satisfies the inequality

$$p(nr, n + j) < r^{nr} \exp \{ -j^2/4nr \} < r^{nr} \exp \{ -(n\epsilon)^2/4nr \}$$
$$= r^{nr} \exp \{ -n\epsilon^2/4r \}.$$

Thus the number of blocks X_{nr} satisfying (8.8) is at most

$$(nr)r^{nr} \exp \{ -n\epsilon^2/4r \} = (nr)r^{nr}(1 + c_1)^{-n}$$

where c_1 is defined by the relation $1 + c_1 = \exp \{ \epsilon^2/4r \}$.

LEMMA 8.8. *For all sufficiently large m the number of blocks X_m satisfying*

$$(8.9) \qquad \left| N(b, X_m) - \frac{m}{r} \right| > m\epsilon$$

is less than $mr^m(1 + c)^{-m}$, where c is a positive constant not dependent on m.

Proof. If m is a multiple of n, this lemma is implied by Lemma 8.7. Thus our problem is to extend Lemma 8.7 for say $m = nr + d$ with $0 \leq d < r$. To any block X_{nr} there correspond r^d blocks X_m where the first nr digits of X_m coincide with X_{nr}, and the last d digits of X_m are arbitrary.

We now prove that, given any block of digits X_m which satisfies inequality (8.9), then the first nr digits of this block, X_{nr}, satisfy (8.8). This is almost intuitively clear since the right side of (8.8), namely $n\epsilon$, is replaced in (8.9) by the much larger value $m\epsilon$. To spell out the details, we assume that a certain X_m satisfies (8.9). Then we use the triangle inequality to write

$$m\epsilon < \left| N(b, X_m) - \frac{m}{r} \right| \leq |N(b, X_m) - N(b, X_{nr})| +$$
$$|N(b, X_{nr}) - n| + \left| n - \frac{m}{r} \right|$$
$$\leq d + |N(b, X_{nr}) - n| + 1,$$

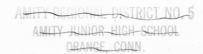

whence we have, for sufficiently large n,

$$|N(b, X_{nr}) - n| > m\epsilon - d - 1 \geqq nr\epsilon - d - 1 > n\epsilon,$$

and this is inequality (8.8).

Hence we can prove the present lemma by using the estimate from the previous lemma, multiplying by r^d to account for the extra digits in X_m. This gives the bound

$$r^d(nr)r^{nr}(1 + c_1)^{-n} \leqq mr^m\{(1 + c_1)^{n/m}\}^{-m}$$
$$< mr^m\{(1 + c_1)^{1/2r}\}^{-m} = mr^m(1 + c)^{-m}.$$

Thus far the notation X_m has designated the block of digits $x_1 x_2 \cdots x_m$. It will be convenient also to regard X_m as the terminating decimal $.x_1 x_2 \cdots x_m$ or the corresponding point in the unit interval.

LEMMA 8.9. *Let ϵ and ϵ_1 be arbitrary positive numbers. Cover each $X_j = .x_1 x_2 \cdots x_j$ for which*

$$\left| N(b, X_j) - \frac{j}{r} \right| > j\epsilon$$

with a closed interval $(X_j, X_j + r^{-j})$ for all $j \geqq m$ and all values of b. For all sufficiently large m the total lengths of these intervals is less than ϵ_1.

Proof. The length of the interval $(X_j, X_j + r^{-j})$ is obviously r^{-j}. We use the bound given in Lemma 8.8 for the number of X_j involved, and also we multiply by r to account for all values of b. Hence the total lengths of the intervals is less than

$$\sum_{j=m}^{\infty} rj(1 + c)^{-j}.$$

This series converges for any value of m, and so we can choose m sufficiently large so that the tail of the series so obtained is arbitrarily small, in fact less than ϵ_1.

THEOREM 8.10. *The set* S *of infinite decimals* $x =$ $.x_1x_2x_3 \cdots$ *which are not simply normal to base* r *has measure zero.*

Proof. Let $T(\epsilon)$ denote the set of those x such that

$$\left| \frac{1}{j} N(b, X_j) - \frac{1}{r} \right| > \epsilon$$

for some b and for infinitely many j. This is the inequality used in Lemma 8.9, and x is in the interval $(X_j, X_j + r^{-j})$. Hence, by Lemma 8.9, the set $T(\epsilon)$ has measure zero. By (8.1) we note that $T(\epsilon)$ is a subset of S. In fact we have for any positive ϵ

$$S = T(\epsilon) \cup T(\epsilon/2) \cup T(\epsilon/4) \cup \cdots.$$

Each set in the union on the right has measure zero, and so S has measure zero by Lemma 8.3.

THEOREM 8.11. *Almost all real numbers are normal to every base.*

Proof. By Lemma 8.1 this can be established by proving that almost all real numbers x have the property that, for every $r > 1$, each of $x, rx, r^2x, \cdots$ is simply normal to all of the bases $r, r^2, r^3, \cdots$. This is equivalent to proving that the following set has measure zero: the set of real numbers x such that r^mx is not simply normal to base r^n for some positive integers $r \geq 2$, $m \geq 0$, $n \geq 1$. The proof is by a succession of applications of Lemma 8.3.

Let $S(r)$ denote the set of those real numbers that are not simply normal to base r. That portion of the set $S(r)$ which lies in the unit interval was designated by S in Theorem 8.10. Now by that theorem S has measure zero, and so the set $S(r)$ has measure zero. This applies to any base, so the set $S(r^n)$ has measure zero for any r and n. Let $V(r)$ denote the union of the sets $S(r^n)$ for $n = 1, 2, 3, \cdots$, and thus $V(r)$ has measure zero. If each element of $V(r)$ is multiplied by r^{-m}, there results a set which we denote by $r^{-m} V(r)$. This set has measure zero because any collection of intervals covering $V(r)$ can be transformed by similar use of the multiplicative factor r^{-m} to give a covering of $r^{-m} V(r)$. Now $r^{-m} V(r)$ consists of those real numbers x such that $r^m x$ is not simply normal to base r^n for at least one value of n. If $W(r)$ denotes the union of $r^{-m} V(r)$ for $m = 0, 1, 2, \cdots$, and W the union of $W(r)$ for $r = 2, 3, 4, \cdots$, then W also has measure zero, and the proof is complete.

3. Equivalent definitions. We return to the various definitions of normal number given in § 1, and prove their equivalence.

LEMMA 8.12. *Let r, k, and s be given positive integers with $r \geqq 2$, $1 \leqq s \leqq k$. Let B_k be a fixed block of k digits to base r and ϵ any positive real number. Then for all t sufficiently large the inequality*

$$(8.10) \qquad N_s(B_k, A_t) > \frac{t}{kr^k} - \frac{2\epsilon t}{k}$$

is satisfied by all blocks A_t of t digits apart from at most ϵr^t exceptional blocks.

Proof. First we reformulate Lemma 8.8 with the base r replaced by r^k. Thus the number of blocks X_m of m digits to base r^k satisfying

$$\left| N(b, X_m) - \frac{m}{r^k} \right| > m\epsilon$$

is less than $mr^{km}(1 + c)^{-m}$, where the positive constant c is independent of m. Here b is a single digit to base r^k. To say this another way, the inequality

$$m\epsilon \geqq N(b, X_m) - \frac{m}{r^k} \geqq -m\epsilon$$

is satisfied by all blocks X_m apart from at most

$$mr^{km}(1 + c)^{-m}$$

special blocks. A fortiori, we can say that the inequality

$$(8.11) \qquad N(b, X_m) \geqq \frac{m}{r^k} - m\epsilon$$

holds for all X_m apart from at most $mr^{km}(1 + c)^{-m}$ blocks. We now interpret this in base r, so that b becomes a block B_k of k digits and X_m becomes say X_{mk} with mk digits. Also $N(b, X_m)$ becomes $N_1(B_k, X_{mk})$, the subscript 1 occurring now because we do not count the total number of occurrences of B_k in X_{mk}, but only those appropriately located. We can extend this to get

$$(8.12) \qquad N_1(B_k, X_{mk}) = N_s(B_k, A_t)$$

by regarding A_t as the block X_{mk} with $s - 1$ digits attached at the left end, and $t - mk - (s - 1)$ digits attached at the right end. We can be certain that (8.12) holds if the number of digits attached at the right end is fewer than k. This we do by regarding the integers t, s, and k as given, and from them we determine m as the quotient when we apply the division algorithm to $t - s + 1$ and k,

$$t - s + 1 = mk + u, \qquad 0 \leqq u < k.$$

This implies that

$$(8.13) \quad mk \leqq t < mk + 2k, \qquad \frac{t}{k} \geqq m > \frac{t}{k} - 2.$$

Thus we have obtained (8.12) by identifying most of the digits of A_t with the block X_{mk},

$$A_t = a_1 a_2 \cdots a_t$$

$$= a_1 a_2 \cdots a_{s-1} x_1 x_2 \cdots x_{mk} a_{t-u+1} a_{t-u+2} \cdots a_t,$$

with fewer than k digits both prior to and following X_{mk} because $s - 1 < k$ and $u < k$. Corresponding to a fixed set of values for X_{mk} there are r^{t-mk} blocks A_t because of the $t - mk$ digits available at the two ends.

Thus we can rewrite (8.11) in view of (8.12) and the other discussion, and conclude that for all sufficiently large t the inequality

$$(8.14) \qquad N_s(B_k, A_t) \geqq \frac{m}{r^k} - m\epsilon$$

holds for all blocks A_t apart from at most

$$mr^{km}(1 + c)^{-m}r^{t-km} = mr^t(1 + c)^{-m}$$

exceptional blocks. As t increases indefinitely, so does m, and so $m(1 + c)^{-m}$ tends to zero. Thus (8.14) holds for all A_t apart from at most ϵr^t exceptional blocks, provided t is sufficiently large.

Also we note that $\epsilon t/k > 2/r^k$ for t sufficiently large, and so by (8.13)

$$\frac{m}{r^k} - m\epsilon > \left(\frac{t}{k} - 2\right)\frac{1}{r^k} - \left(\frac{t}{k}\right)\epsilon$$

$$= \frac{t}{kr^k} - \frac{t\epsilon}{k} - \frac{2}{r^k} > \frac{t}{kr^k} - \frac{2\epsilon t}{k}.$$

In view of this, we see that the inequality (8.10) is satisfied by any block A_t which satisfies (8.14), and so the lemma is proved.

THEOREM 8.13. *If the infinite decimal* $x = .x_1x_2 \cdots$ *satisfies* (8.2), *then* x *satisfies* (8.4).

Proof. To prove this converse of Lemma 8.1, we begin by observing that (8.2) implies that for any block of t digits A_t and any positive ϵ

$$(8.15) \qquad N(A_t, X_n) > \frac{n}{r^t} - \frac{\epsilon n}{r^t} = \frac{n(1 - \epsilon)}{r^t}$$

for all sufficiently large n.

Recall that $N_s(B_k, X_n)$ counts the number of occurrences of B_k in X_n of a certain type, namely $b_1 = x_m$, $b_2 = x_{m+1}$, $\cdots$, $b_k = x_{m+k-1}$ with $m \equiv s \pmod{k}$. We establish that, for $t > k$,

$$(8.16)$$
$$(t - k + 1)N_s(B_k, X_n) \geqq \sum_{j=1}^{n-t+1} N_s(B_k, x_j x_{j+1} \cdots x_{j+t-1}),$$

where the notation N_s in the sum on the right is to be interpreted as counting occurrences of B_k subject to the same condition on the subscripts: namely, occurrences of the form $b_1 = x_m$, $\cdots$, $b_k = x_{m+k-1}$ with $m \equiv s \pmod{k}$. Any such occurrence of B_k in X_n is counted exactly $t - k + 1$ times by the expression on the left side of (8.16), and at most $t - k + 1$ times by the expression on the right as j ranges over its assigned values.

Having established (8.16), we now approximate the sum on the right side of this inequality. As j ranges over the values $1, 2, \cdots, n - t + 1$, the block $x_j x_{j+1} \cdots x_{j+t-1}$ is identical with any specified block A_t of t digits at least $nr^{-t}(1 - \epsilon)$ times, by (8.15). Hence we can change the sum in (8.16) to a sum over all the r^t possible values A_t, thus

$$(t - k + 1)N_s(B_k, X_n) > nr^{-t}(1 - \epsilon) \sum_{A_t} N_\sigma(B_k, A_t).$$

This sum ranges over all possible blocks A_t, and the indeterminate subscript σ can remain entirely unspecified because we now apply Lemma 8.12. Thus we apply the inequality (8.10) to $N_\sigma(B_k, A_t)$ in all cases of nonexceptional blocks A_t, in number at least $r^t - \epsilon r^t$. Disregarding the exceptional blocks, we obtain

$$(t - k + 1)N_s(B_k, X_n)$$

$$> nr^{-t}(1 - \epsilon)(r^t - \epsilon r^t)\left(\frac{t}{kr^k} - \frac{2\epsilon t}{k}\right)$$

$$= nt\left(\frac{1}{kr^k} - \frac{2\epsilon}{k}\right)(1 - \epsilon)^2$$

$$> nt\left(\frac{1}{kr^k} - 2\epsilon\right)(1 - 2\epsilon)$$

$$> nt\left(\frac{1}{kr^k} - 4\epsilon\right)$$

for n and t sufficiently large. (The steps in this chain of inequalities are incorrect in case ϵ is large, for example $\epsilon = 1$, but the final result is obviously correct for such values of ϵ.) Hence we have, for any positive ϵ,

$$\frac{N_s(B_k, X_n)}{n} > \frac{t}{t - k + 1}\left(\frac{1}{kr^k} - 4\epsilon\right) > \frac{1}{kr^k} - 4\epsilon,$$

$$\lim_{n \to \infty} \inf \frac{N_s(B_k, X_n)}{n} \geqq \frac{1}{kr^k}.$$

This result holds for the r^k possible blocks B_k and the k possible values of s. Hence we can apply Lemma 8.2 to obtain (8.4), and the proof of the theorem is complete.

THEOREM 8.14. *The definitions (8.4) and (8.5) are equivalent to (8.2).*

Proof. Theorem 8.13 established that (8.2) implies (8.4). It is obvious that (8.4) implies (8.5). The logic will be complete if we prove that (8.5) implies (8.2), which we now do. By (8.5) we see that, for any block of digits A_t and any positive ϵ,

$$(8.17) \qquad N_1(A_t, X_n) > \frac{n}{tr^t} - \frac{\epsilon n}{tr^t}$$

for all n sufficiently large.

Next we establish that, with $k < t < n$,

$$(8.18) \quad N(B_k, X_n) \geqq \sum_{A_t} N(B_k, A_t)N_1(A_t, X_n)$$
$$> \left(\frac{n}{tr^t} - \frac{\epsilon n}{tr^t}\right) \sum_{A_t} N(B_k, A_t),$$

the second inequality arising from (8.17). This is seen by partitioning X_n into consecutive blocks of t digits and then counting $N_1(A_t, X_n)$. Then we count $N(B_k, A_t)$, the number of occurrences of B_k in A_t, perform the obvious multiplication, and sum over all possible A_t. This sum may not yield exactly $N(B_k, X_n)$, because we lose any B_k that straddles blocks of length t in the partitioning of X_n.

The expression $N(B_k, A_t)$ on the right side of (8.18) can be written

$$N(B_k, A_t) = \sum_{s=1}^{k} N_s(B_k, A_t),$$

and Lemma 8.12 is applied to these terms. Thus (8.18) implies

$$N(B_k, X_n) > \left(\frac{n}{tr^t} - \frac{\epsilon n}{tr^t}\right)(r^t - \epsilon r^t)k\left(\frac{t}{kr^k} - \frac{2\epsilon t}{k}\right)$$
$$= n\left(\frac{1}{r^k} - 2\epsilon\right)(1 - \epsilon)^2$$
$$> n(r^{-k} - 2\epsilon)(1 - 2\epsilon)$$
$$> n(r^{-k} - 4\epsilon),$$

for all n sufficiently large. (As in the proof of the previous theorem, the steps in this chain of inequalities are incorrect if ϵ is large, but the final result holds nevertheless.) Thus we have

$$\liminf_{n \to \infty} \frac{N(B_k, X_n)}{n} \geqq \frac{1}{r^k}$$

for any block B_k of k digits. Application of Lemma 8.2 leads to the conclusion (8.2), which proves the theorem.

There is a connection between normal numbers and the concept of uniform distribution modulo 1, the definition of which was given in § 3 of Chapter 6.

THEOREM 8.15. *The number x is normal to base r if and only if the numbers x, rx, r^2x, $\cdots$ are uniformly distributed modulo 1.*

Proof. Suppose first that x, rx, r^2x, $\cdots$ are uniformly distributed modulo 1. Let the decimal expansion of the fractional part of x be $.x_1x_2x_3 \cdots$ to base r, and let $B_k = b_1b_2 \cdots b_k$ be any block of k digits. We must prove that B_k occurs in $.x_1x_2x_3 \cdots$ with frequency r^{-k}. Let I denote the open interval of points y satisfying

$$.b_1b_2 \cdots b_k < y < r^{-k} + .b_1b_2 \cdots b_k,$$

so that I has length r^{-k}. The decimal expansion to base r of every such number y begins with the digits $b_1b_2 \cdots b_k$: that is, B_k.

Next, if the fractional part (r^mx) of any number r^mx has decimal expansion beginning with $b_1b_2 \cdots b_k$, then (r^mx) belongs to the interval I. For the only other possibilities are that (r^mx) is one of the end points of the closure of the interval, such as $(r^mx) = .b_1b_2 \cdots b_k$. But these two possibilities are clearly ruled out by the hypothesis. It follows from the definition of uniform distribution that

$$\lim_{n \to \infty} \frac{1}{n} N(B_k, X_n) = \lim_{n \to \infty} \frac{n(I)}{n} = r^{-k},$$

where $n(I)$ denotes the number of those points (x), (rx), $\cdots$, $(r^{n-1}x)$ which lie in the interval I.

Conversely, suppose that x is normal to base r. For any positive integer m, divide the unit interval into r^m closed subintervals,

(8.19)

$$(0, r^{-m}), (r^{-m}, 2r^{-m}), 2r^{-m}, 3r^{-m}), \cdots, (1 - r^{-m}, 1).$$

Note that $(r^j x)$, being irrational for every integer j, is not the end point of any of these intervals. The normality of the number x implies that the points (x), (rx), $(r^2 x)$, $\cdots$ are distributed with equal frequency in these intervals. That is to say, if R denotes any one of the subintervals (8.19), and if $n(R)$ denotes the number of those points (x), (rx), $\cdots$, $(r^{n-1}x)$ which lie in R, then

(8.20)
$$\lim_{n \to \infty} \frac{n(R)}{n} = r^{-m}.$$

Next, let I be any subinterval of the unit interval, say $I = (\alpha, \beta)$ with $0 \leqq \alpha < \beta \leqq 1$. We are not assuming anything about the closure properties of I, so that α and β may or may not belong to I. Let R_1 be the collection of those intervals (8.19) that lie entirely inside I. Thus the length of R_1 is at least $\beta - \alpha - 2r^{-m}$. Hence, given any $\epsilon > 0$, we see by (8.20) that, for all n sufficiently large,

$$\frac{n(I)}{n} \geqq \frac{n(R_1)}{n} \geqq \beta - \alpha - 2r^{-m} - \frac{\epsilon}{2}.$$

If we choose m large enough so that $r^{-m} < \epsilon/4$, then we can conclude that

$$\frac{n(I)}{n} \geqq \beta - \alpha - \epsilon$$

for all n sufficiently large.

Similarly, if we write R_2 for the collection of those intervals (8.19) that have one or more points in common with I, we see that the length of R_2 is less than $\beta - \alpha + 2r^{-m}$. Also we conclude that

$$\frac{n(I)}{n} \leqq \frac{n(R_2)}{n} \leqq \beta - \alpha + 2r^{-m} + \frac{\epsilon}{2} \leqq \beta - \alpha + \epsilon$$

for all n sufficiently large. Since ϵ can be made arbitrarily small, we conclude that

$$\lim_{n \to \infty} \frac{n(I)}{n} = \beta - \alpha,$$

and the theorem is proved.

Theorem 8.15 can be used to prove such a proposition as: if x is normal to base r, so is jx for any non-zero integer j. This result holds also for any rational $j \neq 0$, but the proof of this stronger statement is more difficult.

4. A normal number exhibited. It is not known whether such numbers as $\sqrt{2}$, e, and π are normal to any base. It turns out that the decimal expansion obtained by writing the natural numbers in order gives a normal number. For convenience we shall prove this for the special base 10, although the proof is readily extended to any base.

Theorem 8.16. *The number*

$$(8.21) \quad x = 0.1\ 2\ 3\ 4\ 5\ 6\ 7\ 8\ 9\ 10\ 11\ 12\ 13\ 14\ \cdots$$

formed by writing the natural numbers in succession is normal to base 10.

Proof. By X_n we shall mean, as usual, the block consisting of the first n digits of x in (8.21). We think of X_n as partitioned into blocks corresponding to the natural numbers; thus

$$(8.22) \quad X_n = 1, 2, 3, 4, 5, 6, 7, 8, 9, 10,$$
$$11, 12, 13, 14, \cdots, a_1 a_2 \cdots a_m, \cdots.$$

The last complete natural number in this partitioning of X_n is assumed to have digits $a_1 a_2 a_3 \cdots a_m$, and, if we denote this number by u, we have

$$u = a_1 10^{m-1} + a_2 10^{m-2} + \cdots + a_m, \qquad a_1 \neq 0.$$

In (8.22) there are at most m digits after the last comma, and, since there are at most $u + 1$ partitions, we see that

$$(8.23) \qquad \qquad n \leqq m(u + 1).$$

It is convenient to define the numbers

$$u_j = [u 10^{-j}] = a_1 10^{m-1-j} + a_2 10^{m-2-j} + \cdots + a_{m-j},$$
$$j = 1, 2, \cdots, m - 1.$$

We are concerned with estimating the number of occurrences of B_k in X_n, where $B_k = b_1 b_2 \cdots b_k$. In making this estimate we count only the occurrences of B_k inside the partitions which make up (8.22), ignoring occurrences that straddle the commas therein. We look for blocks of digits in (8.22) having the form

$$(8.24) \quad y_1 y_2 \cdots y_s b_1 b_2 \cdots b_k z_1 z_2 \cdots z_t = Y_s B_k Z_t.$$

There being at most m digits in any partition in (8.22), we require that $s + k + t \leqq m$. Moreover the natural number corresponding to the block of digits (8.24) must not exceed u, and this will be guaranteed if we require that

$$y_1 10^{s-1} + y_2 10^{s-2} + \cdots + y_s < a_1 10^{m-k-t-1}$$
$$+ a_2 10^{m-k-t-2} + \cdots + a_{m-k-t} = u_{k+t}.$$

In view of this, we have $u_{k+t} - 1$ possible values for the block Y_s in (8.24), and 10^t possible values for the t digits Z_t. Thus the number of blocks of digits of the form (8.24) to be found in (8.22) is at least

$$\sum_{t=0}^{m-k-1} 10^t(u_{k+t} - 1),$$

where we get this maximum value $m - k - 1$ for t by setting $s = 1$ in the inequality $s + k + t \leqq m$. Now $N(B_k, X_n)$ is the total number of occurrences of B_k in (8.22), and so we have

$$N(B_k, X_n) \geqq \sum_{t=0}^{m-k-1} 10^t(u_{k+t} - 1).$$

By the definition of u_j we have $u_j > u10^{-j} - 1$, and so

$$N(B_k, X_n) > \sum_{t=0}^{m-k-1} 10^t(u \cdot 10^{-k-t} - 2)$$

$$= \sum_{t=0}^{m-k-1} u \cdot 10^{-k} - \sum_{t=0}^{m-k-1} 2 \cdot 10^t$$

$$> (m - k)u \cdot 10^{-k} - 10^{m-k}.$$

Dividing the left side by n and the right side by $m(u + 1)$, we see by (8.23) that

$$\frac{1}{n} N(B_k, X_n)$$

$$> \frac{m - k}{m} \cdot \frac{u}{u + 1} 10^{-k} - \frac{10^{m-k}}{u + 1} \cdot \frac{1}{m}$$

$$= 10^{-k} - 10^{-k} \left\{ \frac{1}{u + 1} + \frac{u}{u + 1} \cdot \frac{k}{m} \right\} - \frac{10^{m-k}}{u + 1} \cdot \frac{1}{m}.$$

Note that

$$\frac{10^{m-k}}{u+1} \leqq \frac{10^{m-1}}{u+1} \leqq \frac{u}{u+1} < 1.$$

Also we see that, as n tends to infinity,

$$u \to \infty, \qquad m \to \infty, \qquad \frac{k}{m} \to 0.$$

Hence, for any given $\epsilon > 0$, we can choose n sufficiently large so that

$$\frac{1}{n} N(B_k, X_n) > 10^{-k} - \epsilon,$$

$$\liminf_{n \to \infty} \frac{1}{n} N(B_k, X_n) \geqq 10^{-k}.$$

This holds for all blocks B_k of k digits, and so we apply Lemma 8.2 to get

$$\lim_{n \to \infty} \frac{N(B_k, X_n)}{n} = 10^{-k}.$$

Thus x is normal to base 10.

This proof could be shortened slightly by the use of Lemma 8.12, but we have preferred to give a direct argument independent of that lemma.

Notes on Chapter 8

Normal numbers were introduced, with definition as in equations (8.4), by E. Borel, who proved the central Theorem 8.11 that almost all real numbers are normal; cf. *Rend. Circ. Mat. Palermo*, **27** (1909), 247–271. The special case of Theorem 8.11 that almost all real numbers are simply normal can be proved as an application of the strong law of large numbers in the theory of probability; cf. Paul R. Halmos, *Measure Theory*, Van Nostrand, 1950, p. 206.

The equivalence of definitions (8.4) and (8.5), proved in Theorem 8.14, is due to S. S. Pillai, *Proc. Indian Acad. Sci.*, **A, 12** (1940), 179–184. Theorem 8.13 is due to Niven and Zuckerman, *Pacific J. Math.*, **1** (1951), 103–109; the proof given here is due to J. W. S. Cassels, *Pacific J. Math.*, **2** (1952), 555–557. The proof of Theorem 8.14 was suggested by this paper of Cassels and work of J. E. Maxfield, *Pacific J. Math.*, **2** (1952), 23–24.

J. E. Maxfield has treated normal numbers in the n-dimensional case; cf. *Pacific J. Math.*, **3** (1953), 189–196. Maxfield also points out (p. 195) that the set of numbers simply normal to no base whatever is uncountable. It is easy to establish the somewhat weaker proposition that the set of numbers not simply normal to a specific base r is uncountable: consider the uncountable subset thereof of numbers of the form $.0x_2 0x_4 0x_6 \cdots$, where each x_{2n} is either 0 or 1.

Theorem 8.15 is due to D. D. Wall, *Normal Numbers*, Ph.D. thesis (1949), University of California, Berkeley, California.

Theorem 8.16 is due to D. G. Champernowne, *J. London Math. Soc.*, **8** (1933), 254–260. This result has been generalized by A. H. Copeland and P. Erdös, *Bull. Amer. Math. Soc.*, **52** (1946), 857–860.

One topic that we have not touched upon is the ϵ normality of integers of A. S. Besicovitch, *Math. Zeit.*, **39** (1934), 146–156. See also H. A. Hanson, *Can. J. Math.*, **6** (1954), 477–485.

E. Borel defined an *absolutely normal number* as one that is normal to every base. However, might it be that, if a number is normal to one base, then it is absolutely normal? It would be interesting to have an answer to this question, because it would tell us the depth of the normality concept, whether this concept is associated with the numbers themselves (as are the rational and algebraic concepts) or whether it is dependent upon the particular base of representation used to express the number.

The definition (8.2) of a normal number contains superfluous conditions: it is enough to say that the requirements are satisfied for infinitely many integers k. In fact it is not difficult to prove that, if the conditions are satisfied for some fixed k, they must also hold for $k - 1$. However, it is not enough to require that the equations hold for some finite set of values of k: cf. I. J. Good, *J. London Math. Soc.*, **21** (1946), 167–169; also D. Rees in the same journal, **21** (1946), 169–172.

CHAPTER 9

THE GENERALIZED LINDEMANN
THEOREM

1. Statement of the theorem. Transcendental numbers were first exhibited by Liouville, using a technique which we set forth in Theorem 7.9. Later, Hermite in 1873 proved that e is transcendental (Theorem 2.12), and Lindemann extended the method to π in 1882. The transcendence of e and π are special cases of a more general theorem of Lindemann which is the subject of this chapter. We state two equivalent forms.

THEOREM 9.1. *Given any distinct algebraic numbers α_1, α_2, $\cdots$, α_m, the values e^{α_1}, e^{α_2}, $\cdots$, e^{α_m} are linearly independent over the field of algebraic numbers.*

Alternative statement. *Given any distinct algebraic numbers α_1, α_2, $\cdots$, α_m, the equation*

$$(9.1) \qquad \sum_{j=1}^{m} a_j e^{\alpha_j} = 0$$

is impossible in algebraic numbers a_1, a_2, $\cdots$, a_m not all zero.

The proof of this result is given in the next two sections. It is clear that this theorem implies that e is transcendental, indeed that e^{α} is transcendental for any non-zero algebraic number α. This observation includes the transcendence of π as a special case. For, if π were algebraic, so would $i\pi$ be algebraic by Theorem 7.2, and hence $e^{i\pi}$

would be transcendental, contrary to the well-known fact that $e^{i\pi} = -1$. Theorem 9.1 has other consequences, but we postpone further discussion of these until § 4 in order to get on with the proof of the theorem itself.

2. Preliminaries.   We begin by outlining some needed algebraic background material, in particular some deeper properties of algebraic numbers than we have discussed heretofore. No proofs will be given of Theorems 9.4, 9.5, or of the fund mental result on the algebraic dependence of any symmetric polynomial on the elementary symmetric polynomials. These are standard results; so we merely cite references.

The elementary symmetric functions $\sigma_1, \sigma_2, \cdots, \sigma_n$ of $x_1, x_2, \cdots, x_n$ can be defined by the identity

$$(y - x_1)(y - x_2) \cdots (y - x_n)$$
$$= y^n - \sigma_1 y^{n-1} + \sigma_2 y^{n-2} - \cdots + (-1)^n \sigma_n.$$

In general a symmetric function of $x_1, x_2, \cdots, x_n$ is one that is invariant under any permutation of $x_1, \cdots, x_n$. Any symmetric polynomial in $x_1, \cdots, x_n$ with coefficients in a field F is expressible as a polynomial in the elementary symmetric functions $\sigma_1, \cdots, \sigma_n$ with coefficients in F. Moreover, if the symmetric polynomial in $x_1, \cdots, x_n$ has rational integers for coefficients, so does the polynomial in $\sigma_1, \cdots, \sigma_n$. Proofs of these results can be found readily in books on the theory of equations; for example, J. V. Uspensky, *Theory of Equations*, McGraw-Hill, p. 63, p. 264 (1948). These results imply the following proposition.

THEOREM 9.2.   *Let $\beta_1, \beta_2, \cdots, \beta_n$ be the roots of a polynomial equation*

$$f(x) = bx^n + c_1 x^{n-1} + c_2 x^{n-2} + \cdots + c_n = 0,$$

with rational integral coefficients. Let $P(x_1, x_2, \cdots, x_n)$ be a symmetric polynomial in $x_1, \cdots, x_n$ with rational coeffi-

cients. *Then $P(\beta_1, \beta_2, \cdots, \beta_n)$ is a rational number. Moreover, if the polynomial P has integral coefficients and is of degree t, then $b^t P(\beta_1, \beta_2, \cdots, \beta_n)$ is an integer.*

Proof. The first part follows from the close relation between the elementary symmetric functions of $\beta_1, \cdots, \beta_n$ and the coefficients of $f(x)$: $\sigma_1 = -c_1/b$, etc. Now $b\beta_1, b\beta_2, \cdots, b\beta_n$ are the roots of

$$b^{n-1}f(x/b) = x^n + c_1 x^{n-1} + bc_2 x^{n-2} + \cdots + b^{n-1}c_n = 0,$$

and so the elementary symmetric functions of $b\beta_1, b\beta_2, \cdots, b\beta_n$ are integers. If $p(x_1, x_2, \cdots, x_n)$ is a homogeneous symmetric polynomial of degree $r \leq t$ with rational integral coefficients, then

$$b^r p(\beta_1, \beta_2, \cdots, \beta_n) = p(b\beta_1, b\beta_2, \cdots, b\beta_n),$$

and so $b^t p(\beta_1, \beta_2, \cdots, \beta_n)$ is an integer. By separating the polynomial P of the theorem into a sum of homogeneous polynomials p, we readily extend this argument from p to P.

We shall also make use of the following result on symmetric functions.

LEMMA 9.3. *Consider the q polynomials $P_1, P_2, \cdots, P_q$ in $y_1, y_2, \cdots, y_m$,*

$$P_j = f_1(x_j)y_1 + f_2(x_j)y_2 + \cdots + f_m(x_j)y_m,$$

$$j = 1, 2, \cdots, q,$$

with coefficients $f_i(x_j)$, where all the $f_i(x)$ are polynomials over any field F. The product of these polynomials, the terms in y being collected, has coefficients which are symmetric polynomials in $x_1, x_2, \cdots, x_q$.

Proof. Let us write the product

$$(9.2) \qquad P_1 P_2 \cdots P_q = \sum_{\substack{i_j = 1 \\ i_1 \leq i_2 \leq \cdots \leq i_q}}^{m} c y_{i_1} y_{i_2} \cdots y_{i_q}.$$

The condition $i_1 \leqq i_2 \leqq \cdots \leqq i_q$ is imposed on the sum to indicate that the terms have been collected. We must prove that each coefficient $c = c(x_1, \cdots, x_q)$ is symmetric in the $x_1, \cdots, x_q$. Any permutation of $x_1, \cdots, x_q$ applied to (9.2) leaves the left side invariant because it merely permutes the polynomials $P_1, P_2, \cdots, P_q$. Hence such a permutation leaves the right side invariant, and so must leave each coefficient c invariant.

We shall assume that the reader is familiar with the concept of extension of an algebraic number field, as given, for example, in the Carus Monograph by Harry Pollard, *The Theory of Algebraic Numbers*, Chapter IV. In particular, we state without proof the following results, which are Theorems 4.6 and 4.7 of Pollard.

THEOREM 9.4. *If F is an algebraic number field, and if θ is algebraic over F, then any element β of the extension field $F(\theta)$ can be expressed as a unique polynomial in θ with coefficients in F,*

$$\beta = a_0 + a_1\theta + a_2\theta^2 + \cdots + a_{n-1}\theta^{n-1},$$

where n is the degree of θ over F.

We shall need a generalization of this to a multiple extension $F(\theta_1, \theta_2, \cdots, \theta_s)$, that any element of this extension is expressible as a polynomial in $\theta_1, \theta_2, \cdots, \theta_s$ with coefficients in F. This generalization is immediately obtained from Theorem 9.4 by use of mathematical induction on s.

THEOREM 9.5. *A multiple algebraic extension of an algebraic number field F is a simple algebraic extension. In other words, if the numbers $\alpha_1, \alpha_2, \cdots, \alpha_s$ are algebraic over F, then there is an algebraic number γ such that the extension of F by γ is the same as the extension of F by $\alpha_1, \alpha_2, \cdots, \alpha_s$; thus*

$$F(\gamma) = F(\alpha_1, \alpha_2, \cdots, \alpha_s).$$

We shall write R for the field of rational numbers. The algebraic field $R(\theta)$ is said to be *normal* over R, provided that any polynomial irreducible over R that has one root in $R(\theta)$ has all roots in $R(\theta)$.

THEOREM 9.6. *Given any algebraic numbers α_1, α_2, $\cdots$, α_s, there exists an algebraic number θ such that the extension field $R(\theta)$ contains the field $R(\alpha_1, \cdots, \alpha_s)$, and $R(\theta)$ is normal.*

Proof. First we apply Theorem 9.5 to obtain γ such that $R(\gamma) = R(\alpha_1, \cdots, \alpha_s)$. Now let the minimal equation of γ be $h(x) = 0$ with coefficients in R, and let the roots of this be $\gamma = \gamma_1, \gamma_2, \cdots, \gamma_m$. Again we apply Theorem 9.5 to find an algebraic number θ such that

$$R(\theta) = R(\gamma_1, \gamma_2, \cdots, \gamma_m) \supset R(\gamma_1) = R(\alpha_1, \alpha_2, \cdots, \alpha_s).$$

To prove that $R(\theta)$ is normal, let $g(x) = 0$ be an irreducible equation over R having a root ρ in $R(\theta)$. Thus $g(x)$ is the minimal polynomial of ρ. Now, by the generalization of Theorem 9.4, ρ is a polynomial in $\gamma_1, \gamma_2, \cdots, \gamma_m$ with rational coefficients, say $\rho = f(\gamma_1, \gamma_2, \cdots, \gamma_m)$. We form the polynomial

$$G(x) = \Pi\{x - f(\gamma_{i_1}, \gamma_{i_2}, \cdots, \gamma_{i_m})\}$$

of degree $m!$ in x, where the product is taken over all permutations $i_1, i_2, \cdots, i_m$ of $1, 2, \cdots, m$. The coefficients of $G(x)$ are symmetric polynomials in its roots: that is, in $f(\gamma_{i_1}, \gamma_{i_2}, \cdots, \gamma_{i_m})$. Any permutation of $\gamma_1, \cdots, \gamma_m$ merely permutes the $f(\gamma_{i_1}, \gamma_{i_2}, \cdots, \gamma_{i_m})$ among themselves. Hence the coefficients of $G(x)$ are symmetric polynomials in $\gamma_1, \gamma_2, \cdots, \gamma_m$, and so are rational numbers by Lemma 9.2. The polynomials $g(x)$ and $G(x)$ have the root ρ in common, and hence the minimal polynomial $g(x)$ is a factor of $G(x)$. But all roots of $G(x) = 0$ are elements of the field $R(\theta)$, and so all roots of $g(x) = 0$ are in $R(\theta)$, and the proof is complete.

Let the degree of the normal field $R(\theta)$ be n, so that θ satisfies a minimal equation with rational coefficients, say

$$(9.3) \quad f(x) = x^n + b_1 x^{n-1} + b_2 x^{n-2} + \cdots + b_n = 0,$$

irreducible over R. Any element of $R(\theta)$ is expressible as a unique polynomial in θ of degree at most $n - 1$. Write $\theta = \theta^{(1)}, \theta^{(2)}, \theta^{(3)}, \cdots, \theta^{(n)}$ for the roots of (9.3). Since $R(\theta)$ is normal over R, all these roots belong to $R(\theta)$; so we can write these *conjugates* in the form

$$(9.4) \qquad \theta^{(j)} = h_j(\theta), \qquad j = 1, 2, \cdots, n,$$

where the $h_j(\theta)$ are polynomials in θ with coefficients in R. The elementary symmetric functions of $h_1(\theta)$, $h_2(\theta)$, $\cdots$, $h_n(\theta)$ are also polynomials in θ with coefficients in R, and so these reduce to the rational numbers $-b_1$, b_2, $-b_3$, $\cdots$, $(-1)^n b_n$ of (9.3) when the relation

$$\theta^n = -b_1 \theta^{n-1} - b_2 \theta^{n-2} - \cdots - b_n$$

is used to eliminate θ^n and higher powers.

Now $\theta^{(2)}$ satisfies the same relation, and therefore the elementary symmetric functions of $h_1(\theta^{(2)})$, $h_2(\theta^{(2)})$, $\cdots$, $h_n(\theta^{(2)})$ are also equal to $-b_1$, b_2, $-b_3$, $\cdots$, $(-1)^n b_n$. Hence we conclude that the numbers

$$h_1(\theta^{(2)}), \, h_2(\theta^{(2)}), \, \cdots, \, h_n(\theta^{(2)})$$

are the roots of (9.3), and so are the same as $\theta^{(1)}$, $\theta^{(2)}$, $\theta^{(3)}$, $\cdots$, $\theta^{(n)}$ in some order. Another way of stating this is that, if $\theta^{(1)}$, $\theta^{(2)}$, $\cdots$, $\theta^{(n)}$ are regarded as polynomials in θ, and if in these polynomials θ is replaced by $\theta^{(2)}$, there merely results a permutation of the conjugates $\theta^{(1)}$, $\cdots$, $\theta^{(n)}$. There is nothing special about $\theta^{(2)}$ in this analysis; so we can generalize to any $\theta^{(i)}$.

LEMMA 9.7. *Let $R(\theta)$ be a normal algebraic extension of degree n over R, the conjugates of θ being $\theta = \theta^{(1)}, \theta^{(2)}, \cdots, \theta^{(n)}$. These conjugates, regarded as polynomials in θ, are merely permuted by the substitution of $\theta^{(i)}$ for θ. More generally, if $F(x)$ is any rational polynomial, then the set*

$$(9.5) \qquad F(\theta^{(1)}), F(\theta^{(2)}), \cdots, F(\theta^{(n)})$$

is permuted by the substitution of any $\theta^{(i)}$ for θ.

Proof. The set (9.5) can be regarded as a set of polynomials in θ by virtue of (9.4). Replacement of θ by $\theta^{(i)}$ permutes the conjugates $\theta^{(1)}, \cdots, \theta^{(n)}$ by the argument preceding the lemma, and so permutes the values (9.5) in the same way.

Any element γ of $R(\theta)$ is a polynomial in θ with rational coefficients, say $\gamma = F(\theta)$, and the values (9.5) are called the conjugates of γ over the field $R(\theta)$. These conjugates are also written as $\gamma = \gamma^{(1)}, \gamma^{(2)}, \cdots, \gamma^{(n)}$. The elementary symmetric polynomials of $\gamma^{(1)}, \cdots, \gamma^{(n)}$ are symmetric polynomials in $\theta^{(1)}, \cdots, \theta^{(n)}$, and so are rational numbers. This proves the following result.

LEMMA 9.8. *Any element γ of $R(\theta)$ and its conjugates over $R(\theta)$ satisfy a polynomial equation $g(x) = 0$ of degree n with integral coefficients.*

We conclude this section with the following rather special result.

LEMMA 9.9. *Consider the functions*

$$f(x) = \sum_{j=1}^{m} a_j x^{\alpha_j}, \qquad g(x) = \sum_{j=1}^{t} b_j x^{\beta_j},$$

with non-zero complex coefficients a_j and b_j, and exponents α_j and β_j that are algebraic numbers. Assume that the α_j are distinct, and likewise that the β_j are distinct. If the product $f(x) \, g(x)$ is formed, and all terms with equal expo-

nents are combined, then there is at least one non-zero coefficient in the result.

Proof. Applying Theorem 9.6 to the numbers $\alpha_1, \cdots,$
$\alpha_m, \beta_1, \cdots, \beta_t,$ we get a normal field $R(\theta)$ containing all
these numbers. As before, let n denote the degree of the
field $R(\theta)$. By Theorem 9.4 we can write each α_j as a
unique polynomial in θ with rational coefficients,

$$\alpha_j = \sum_{i=0}^{n-1} r_{ji}\theta^i.$$

We order these α_j as follows: we say that α_j precedes α_k
in case the first non-zero term of

$$r_{j0} - r_{k0}, r_{j1} - r_{k1}, r_{j2} - r_{k2}, \cdots$$

is positive. In terms of this ordering principle, let us rearrange notation so that α_1 is first among the α_j, and β_1
first among the β_j. Then it is clear that $\alpha_1 + \beta_1$ is first
among all sums $\alpha_j + \beta_k$. Thus, in the product $f(x)\,g(x)$,
the term $a_1b_1x^{\alpha_1+\beta_1}$ has a unique exponent, and so cannot
be combined with or canceled by any other terms.

3. Proof of the theorem. In proving the generalized
Lindemann Theorem 9.1, we find it convenient first to
establish the following special case.

THEOREM 9.10. *Given any m distinct algebraic numbers*
$\alpha_1, \alpha_2, \cdots, \alpha_m,$ *the values* $e^{\alpha_1}, e^{\alpha_2}, \cdots, e^{\alpha_m}$ *are linearly inde-*
pendent over the field of R of rational numbers.

Proof. Let us assume, contrary to the theorem, that
there is a relation

$$(9.6) \qquad \sum_{j=1}^{m} a_j \exp{(\alpha_j)} = 0$$

with rational coefficients not all zero. Discarding terms
with zero coefficients and rearranging notation, we may

presume that no coefficient is zero. Furthermore, by multiplying (9.6) by a suitable integer, we may presume that the coefficients are non-zero rational integers. We use Theorem 9.6 to get a normal field $R(\theta)$ which contains $\alpha_1, \alpha_2, \cdots, \alpha_m$. Let n be the degree of $R(\theta)$ over R, so that each α_j is expressible as a unique polynomial in θ of degree $n - 1$ with rational coefficients, say

$$\alpha_j = \sum_{i=0}^{n-1} r_{ji}\theta^i, \qquad j = 1, 2, \cdots, m.$$

As in the previous section, let $\theta^{(1)} = \theta,\ \theta^{(2)},\ \cdots,\ \theta^{(n)}$ be the conjugates of θ, and so the conjugates of the α_j over $R(\theta)$ are

$$\alpha_j^{(k)} = \sum_{i=0}^{n-1} r_{ji}(\theta^{(k)})^i,$$
$$j = 1, 2, \cdots, m, \qquad k = 1, 2, \cdots, n.$$

Each $\theta^{(k)}$ is, like θ itself, an algebraic number of degree n; so these polynomial expressions are unique in $\theta^{(k)}$. Consequently the hypothesis that the α_j are distinct implies that for fixed k the $\alpha_j^{(k)}$ are distinct.

Next we form the product

$$(9.7) \qquad 0 = \prod_{k=1}^{n} \sum_{j=1}^{m} a_j \exp\,(\alpha_j^{(k)}) = \sum_{j=0}^{r} c_j \exp\,(\beta_j),$$

and this product vanishes because of (9.6) and the fact that $\alpha_j^{(1)}$ is merely alternative notation for α_j. The right member of (9.7) is the result of multiplying out the product and collecting terms with identical exponents, if any. Thus the β_j are distinct. Since the a_j are rational integers, so are the c_j. Furthermore, since the a_j are non-vanishing, by Lemma 9.9 extended to a k-fold product we can conclude that at least one coefficient c_j does not vanish, say $c_0 \neq 0$.

For a fixed value of j, the n conjugates $\alpha_j^{(k)}$ are permuted by substitution of $\theta^{(i)}$ for θ according to Lemma 9.7. Also by the proof of that lemma the permutation is the same whatever the value of j. Hence the replacement of θ by $\theta^{(i)}$ merely permutes the factors in the product in (9.7), so that the total product is invariant. As to the right member of (9.7), the replacement of θ by $\theta^{(i)}$ replaces each β_j by the conjugate $\beta_j^{(i)}$. Hence (9.7) implies

$$(9.8) \quad 0 = \sum_{j=0}^{r} c_j \exp\left(\beta_j^{(1)}\right) = \sum_{j=0}^{r} c_j \exp\left(\beta_j^{(2)}\right) = \cdots$$

$$= \sum_{j=0}^{r} c_j \exp\left(\beta_j^{(n)}\right).$$

We noted earlier that the $\beta_j^{(1)}$ are distinct. Hence the $\beta_j^{(i)}$ are distinct for fixed i.

We multiply the first sum in (9.8) by $\exp\{-\beta_0^{(1)}\}$, the second sum by $\exp\{-\beta_0^{(2)}\}$, $\cdots$, the last sum by $\exp\{-\beta_0^{(n)}\}$. Define

$$(9.9) \quad \gamma_j^{(i)} = \beta_j^{(i)} - \beta_0^{(i)},$$

$$i = 1, 2, \cdots, n, \qquad j = 1, 2, \cdots, r.$$

Since the $\beta_j^{(i)}$ are distinct for fixed i, no $\gamma_j^{(i)}$ equals zero. Then equations (9.8) can be written

(9.10)

$$0 = c_0 + \sum_{j=1}^{r} c_j \exp\left(\gamma\right)_j^{(1)} = c_0 + \sum_{j=1}^{r} c_j \exp\left(\gamma_j^{(2)}\right) = \cdots$$

$$= c_0 + \sum_{j=1}^{r} c_j \exp\left(\gamma_j^{(n)}\right).$$

Now by Lemma 9.8 the conjugates $\gamma_j^{(1)}, \gamma_j^{(2)}, \cdots, \gamma_j^{(n)}$ are roots of a polynomial with integral coefficients, say

$$(9.11) \quad g_j(z) = b_j z^n + \cdots = b_j \prod_{i=1}^{n} \{z - \gamma_j^{(i)}\},$$

$$j = 1, 2, \cdots, r.$$

We may take the integers $b_j > 0$. Also, since no $\gamma_j^{(i)} = 0$, it follows that the constant terms in these polynomials, $g_j(0)$, are non-zero integers.

This completes the algebraic portion of the proof of Theorem 9.10, and we turn now to its analytic aspect. For any polynomial $f(z)$, define $F(z)$ to be the sum of $f(z)$ and its derivatives,

$$F(z) = f(z) + f'(z) + f^{(2)}(z) + \cdots,$$

so that we have

$$\{F(z) \exp (-z)\}' = -f(z) \exp (-z),$$

$$F(b) - F(0) \exp (b) = - \exp (b) \int_0^b f(z) \exp (-z) \, dz.$$

Substitute for b in this relation the values $\gamma_j^{(i)}$ as in (9.9), multiply each equation so obtained by c_j, and sum over $j = 1, \cdots, r$ and $i = 1, \cdots, n$ to get

$$\sum_{j=1}^{r} \sum_{i=1}^{n} c_j F(\gamma_j^{(i)}) - F(0) \sum_{i=1}^{n} \sum_{j=1}^{r} c_j \exp \{\gamma_j^{(i)}\}$$

$$= - \sum_{j=1}^{r} \sum_{i=1}^{n} c_j \exp \{\gamma_j^{(i)}\} \int_0^{\gamma_j^{(i)}} f(z) \exp (-z) \, dz.$$

Applying (9.10) to the second term of this, we obtain

$$(9.12) \quad \sum_{j=1}^{r} c_j \left\{ \sum_{i=1}^{n} F(\gamma_j^{(i)}) \right\} + n c_0 F(0)$$

$$= - \sum_{j=1}^{r} \sum_{i=1}^{n} c_j \exp \{\gamma_j^{(i)}\} \int_0^{\gamma_j^{(i)}} f(z) \exp (-z) \, dz.$$

We now define the polynomial $f(z)$ as

$$(9.13) \quad f(z) = (b_1 b_2 \cdots b_r)^{prn} z^{p-1} \left\{ \prod_{j=1}^{r} g_j(z) \right\}^p / (p-1)!,$$

where p is a prime to be specified. We shall choose the prime p sufficiently large so that the left side of (9.12) is a non-zero integer whereas the right side is, in absolute value, arbitrarily small. Thus we will have the contradiction which establishes the theorem.

Because of the factor z^{p-1} in $f(z)$, it is clear that

$$0 = f(0) = f'(0) = f^{(2)}(0) = \cdots = f^{(p-2)}(0),$$

$$f^{(p-1)}(0) = (b_1 b_2 \cdots b_r)^{prn} \prod_{j=1}^{r} \{g_j(0)\}^p.$$

We choose $p > b_j$ and $p > g_j(0)$ for $j = 1, 2, \cdots, r$, and hence p is not a divisor of the non-zero integer $f^{(p-1)}(0)$.

On the other hand, for $t \geqq p$ we argue that $f^{(t)}(0)$ is an integer divisible by p. To see this we think of $f(z)$ as a sum of powers of z, so that the coefficient of each term in $f^{(t)}(z)$ has t consecutive integers as factors entering in from the differentiation process. But for $t \geqq p$ the product of t consecutive integers is divisible by $p!$, so that the $(p-1)!$ involved in $f(z)$ is canceled. Moreover we can write

$$(9.14) \qquad f^{(t)}(z) = p(b_1 b_2 \cdots b_r)^{prn} G_t(z),$$

where $G_t(z)$ is a polynomial with integral coefficients of degree at most $prn - 1$. Hence $f^{(t)}(0)$ is an integer divisible by p. Using this and the result of the last paragraph, we conclude that $F(0)$ is an integer prime to p. We require p to satisfy the inequalities $p > n$ and $p > c_0$, so that p is not a divisor of the integer $nc_0 F(0)$ in (9.12).

We now establish that the other term on the left side of (9.12), namely

$$\sum_{j=1}^{r} c_j \left\{ \sum_{i=1}^{n} F(\gamma_j^{(i)}) \right\}$$

is an integer divisible by p. In fact we establish that the inner sum

$$\sum_{i=1}^{n} F(\gamma_j^{(i)}) = \sum_{i=1}^{n} f(\gamma_j^{(i)}) + \sum_{i=1}^{n} f'(\gamma_j^{(i)}) + \sum_{i=1}^{n} f^{(2)}(\gamma_j^{(i)}) + \cdots,$$

is a multiple of p. Since $f(z)$ has the factor $\{g_j(z)\}^p$, we see that (9.11) implies

$$f(\gamma_j^{(i)}) = 0, \quad f'(\gamma_j^{(i)}) = 0, \quad \cdots, \quad f^{(p-1)}(\gamma_j^{(i)}) = 0.$$

As for the higher derivatives, we turn to (9.14), whence

(9.15)

$$\sum_{i=1}^{n} f^{(t)}(\gamma_j^{(i)}) = p \sum_{i=1}^{n} (b_1 b_2 \cdots b_r)^{prn} G_t(\gamma_j^{(i)}), \qquad t \geqq p.$$

Now $G_t(z)$ is of degree at most $prn - 1$, so that in view of the factor b_j^{prn} we can apply Theorem 9.2 to conclude that (9.15) is indeed an integer divisible by p.

We have now established that the two expressions comprising the left side of (9.12) are integers, the first a multiple of p, the second prime to p. Hence the left side of (9.12) is a non-zero integer; so, if we take absolute values in this equation, we can write

$$(9.16) \quad 1 \leqq \left| \sum_{j=1}^{r} \sum_{i=1}^{n} c_j \exp \{\gamma_j^{(i)}\} \int_{0}^{\gamma_j^{(i)}} f(z) \exp(-z) \, dz \right|.$$

Define the following maxima for all i and j:

$$m_1 = \max |c_j|; \qquad m_2 = \max |\exp(\gamma_j^{(i)})|;$$

$$m_3 = \max |\gamma_j^{(i)}|;$$

$m_4 = \max |\exp{(-z)}|$ on the straight line path from

$z = 0$ to $z = \gamma_j^{(i)}$;

$m_5 = \max \left| \prod_{j=1}^{r} g_j(z) \right|$ on the same path.

Note that m_3^{p-1} is the maximum of $|z^{p-1}|$ on the same path. Then the inequality (9.16) implies, with (9.13),

$$1 \leqq rnm_1m_2m_3m_4(b_1b_2 \cdots b_r)^{prn}m_3^{p-1}m_5^p/(p - 1)!$$

$$= rnm_1m_2m_4(b_1^{rn}b_2^{rn} \cdots b_r^{rn}m_3m_5)^p/(p - 1)!.$$

Now $r, n, m_1, m_2, m_3, m_4, m_5, b_1, b_2, \cdots, b_r$ are independent of p, so that the last expression tends to zero as p tends to infinity. Thus we have a contradiction, and the proof of Theorem 9.10 is complete.

It is now an easy matter to establish that the basic Theorem 9.1 is a consequence of Theorem 9.10. Suppose that there is some relation (9.1), where we may assume without loss of generality that none of the algebraic numbers a_j is zero. We use Theorem 9.6 to determine an algebraic field $R(\theta)$, normal over R, which contains the algebraic numbers a_j. The conjugates $a_j^{(i)}$ belong to $R(\theta)$, a field which we presume to have degree q over R. (Heretofore we have used the symbol n to denote the degree of the field $R(\theta)$. Our switch to q is to emphasize that now $R(\theta)$ is an extension field determined by the coefficients a_j, whereas formerly it was determined by the exponents α_j.) We form the product

$$\prod_{i=1}^{q} \{a_1^{(i)} \exp{(\alpha_1)} + a_2^{(i)} \exp{(\alpha_2)} + \cdots + a_m^{(i)} \exp{(\alpha_m)}\} = 0.$$

If we think of the coefficients $a_j^{(i)}$ as polynomials in $\theta^{(i)}$, we see by Lemma 9.3 that this product has coefficients which are symmetric in $\theta^{(1)}$, $\theta^{(2)}$, $\cdots$, $\theta^{(q)}$, and so are rational by Theorem 9.2. By Lemma 9.9 the product does not vanish identically. Thus we have established that a relation (9.1) implies a similar relation with rational coefficients, and this was proved impossible by Theorem 9.10.

4. Applications of the theorem. In § 1 we pointed out that Theorem 9.1 implies the transcendence of e and π. More generally we can state the following.

THEOREM 9.11. *The following numbers are transcendental:*

(a) e, π;

(b) e^{α}, sin α, cos α, tan α, sinh α, cosh α, tanh α *for any non-zero algebraic number α;*

(c) log α, arcsin α, *and in general the inverse functions of those in the list* (b), *for any non-zero algebraic number $\alpha \neq 1$; wherever multiple values are involved, every such value is transcendental.*

Proof. The first three listed, e, π and e^{α}, were treated in § 1. Next, if sin α were algebraic, say sin $\alpha = a$, we would have

$$e^{i\alpha} - e^{-i\alpha} - 2iae^0 = 0,$$

with algebraic coefficients and exponents, contrary to Theorem 9.1. A similar argument applies to the other trigonometric and the hyperbolic functions, each being written in terms of exponential functions.

As for those in group (c) in the theorem, first suppose that log $\alpha = a$ with algebraic a. Then we would have $e^a = \alpha$, and by the earlier argument this is possible only in case $a = 0$, $\alpha = 1$. The other values in group (c) can be treated similarly, the restriction $\alpha \neq 1$ being needed in some cases.

5. Squaring the circle. One of the celebrated problems of antiquity is to construct, by methods of straight-edge and compass, a square equal in area to a given circle. The impossibility of this construction was settled by Lindemann when he proved that π is transcendental. For, on the one hand, all line segments that can be constructed from a given unit length by a finite number of straight-edge and compass constructions have lengths that are *algebraic* numbers. (We do not prove this assertion here, since there are many excellent accounts of such geometric constructions, references being given in the notes at the end of this chapter.) On the other hand, given any circle, we may regard its radius as the unit of length, so that the circle has area π square units. So the problem of constructing a square of equal area is equivalent to the problem of constructing a line of length $\sqrt{\pi}$ from a given unit length. But this is impossible by the admissible procedures, because $\sqrt{\pi}$ is transcendental, since otherwise it would follow from Theorem 7.2 that $\pi = \sqrt{\pi} \cdot \sqrt{\pi}$ is algebraic.

Notes on Chapter 9

The work of C. Hermite on the transcendence of e is in *Compt. Rend. Acad. Sci. Paris*, **77** (1873), 18–24; that of F. Lindemann on the transcendence of π in *Math. Annalen*, **20** (1882), 213–225. This paper of Lindemann is also the source of Theorem 9.1; a more detailed proof was given later by K. Weierstrass, *Math. Werke*, **II** (1895), 341–362. The result is sometimes called the Lindemann–Weierstrass theorem.

In the proof of Theorem 9.1 we have used ideas from two recent papers on the subject: R. Steinberg and R. M. Redheffer, *Pacific J. Math.*, **2** (1952), 231–242; Th. Skolem, *Norske Vid. Selsk. Forhandl.*, *Trondhjem*, **19**, no. 12 (1947), 40–43.

Proofs of the transcendence of e and π are not so difficult as the proof of the more general Theorem 9.1. For example, e was treated in Theorem 2.12 in a fairly simple way. Both e and π are discussed in Hardy and Wright, *The Theory of Numbers*, Chapter XI, by methods avoiding complex integration; π is discussed by Niven, *Amer. Math. Monthly*, **46** (1939), pp. 469–471.

We cite a few of the papers giving applications and extensions of the Lindemann theorem: Verne E. Dietrich and Arthur Rosenthal, *Bull. Amer. Math. Soc.*, **55** (1949), 954–956; William J. LeVeque, *Proc. Amer. Math. Soc.*, **2** (1951), 401–403; R. M. Redheffer, *Amer. Math. Monthly*, **60** (1953), 25–27.

§ 5. Geometric constructions by means of straightedge and compass are discussed in detail in many books, for example the following: Richard Courant and Herbert Robbins, *What Is Mathematics?*, Chapter III, Oxford (1941); H. P. Hudson, *Ruler and Compasses*, Longmans, Green (1916); E. W. Hobson, *Squaring the Circle*, Cambridge (1913); J. W. A. Young, *Monographs on Modern Mathematics*, No. VIII by L. E. Dickson, Longmans, Green (1911).

CHAPTER 10

THE GELFOND–SCHNEIDER THEOREM

1. Hilbert's seventh problem. In 1900 David Hilbert announced a list of twenty-three outstanding unsolved problems. The seventh problem was settled by the publication of the following result in 1934 by A. O. Gelfond, which was followed by an independent proof by Th. Schneider in 1935.

THEOREM 10.1. *If α and β are algebraic numbers with $\alpha \neq 0$, $\alpha \neq 1$, and if β is not a real rational number, then any value of α^β is transcendental.*

Remarks. The hypothesis that "β is not a real rational number" is usually stated in the form "β is irrational." Our wording is an attempt to avoid the suggestion that β must be a real number. Such a number as $\beta = 2 + 3i$, sometimes called a "complex rational number," satisfies the hypotheses of the theorem. Thus the theorem establishes the transcendence of such numbers as 2^i and $2^{\sqrt{2}}$. In general, $\alpha^\beta = \exp\{\beta \log \alpha\}$ is multiple-valued, and this is the reason for the phrase "any value of" in the statement of Theorem 10.1. One value of $i^{-2i} = \exp\{-2i \log i\}$ is e^π, and so this is transcendental according to the theorem.

Before proceeding to the proof of Theorem 10.1, we state an alternative form of the result.

THEOREM 10.2.   *If α and γ are non-zero algebraic numbers, and if $\alpha \neq 1$, then $(\log \gamma)/(\log \alpha)$ is either rational or transcendental.*

To prove that this is an equivalent statement to Theorem 10.1, write β for $(\log \gamma)/(\log \alpha)$, so that $\gamma = \alpha^\beta$. To prove that Theorem 10.1 implies Theorem 10.2, let us assume that the hypotheses of the latter are satisfied, but, contrary to the conclusion, that β is algebraic but not rational. Then γ is transcendental by Theorem 10.1, and we have a contradiction.

Conversely, if we assume the hypotheses of Theorem 10.1 but deny the conclusion thereof, we have an algebraic number α^β which is not zero. From application of Theorem 10.2 we conclude that β is either rational or transcendental, and again we have a contradiction. Thus Theorem 10.2 implies Theorem 10.1.

Theorem 10.2 implies that the common logarithms of positive rational numbers, i.e., logarithms of positive rational numbers to base 10, are either rational or transcendental. This can be seen from the simple relation

$$\log_{10} r = \frac{\log r}{\log 10}.$$

Thus we can state a stronger result than Theorem 2.11: if r is any positive rational number not of the form 10^n where n is a rational integer, then $\log_{10} r$ is transcendental.

2. Background material.   In this section we list the results from algebra and analysis which will be needed in the proof of Theorem 10.1. We state some of the results as lemmas for convenience, giving references but no proofs. Although this prerequisite material is more extensive than any used heretofore in this book, it is very central to the main stream of mathematics. We do need a couple of special results for the proof of the Gelfond–

Schneider theorem, and they will be given with proofs in the next section.

LEMMA 10.3. *Consider a determinant with the non-zero element ρ_j^a in the j-th row and $1 + a$-th column, with $j = 1, 2, \cdots, t$ and $a = 0, 1, \cdots, t - 1$. This is called a Vandermonde determinant, and it vanishes if and only if $\rho_j = \rho_k$ for some distinct pair of subscripts j, k.*

This can be found in J. V. Uspensky, *Theory of Equations*, McGraw-Hill, p. 214. The next four lemmas are in Harry Pollard, *The Theory of Algebraic Numbers*, John Wiley, p. 53, p. 60, pp. 63–66, p. 72.

LEMMA 10.4. *Let α and β be algebraic numbers in a field K of degree h over the rationals. If the conjugates of α for K are $\alpha = \alpha_1, \alpha_2, \cdots, \alpha_h$ and for β are $\beta = \beta_1, \beta_2, \cdots, \beta_h$, then the conjugates of $\alpha\beta$ and $\alpha + \beta$ are $\alpha_1\beta_1, \cdots, \alpha_h\beta_h$ and $\alpha_1 + \beta_1, \cdots, \alpha_h + \beta_h$.*

LEMMA 10.5. *If α is an algebraic number, then there is a positive rational integer r such that $r\alpha$ is an algebraic integer.*

LEMMA 10.6. *If K is an algebraic number field of degree h over the rationals, then there exist integers $\beta_1, \beta_2, \cdots, \beta_h$ in K such that every integer in K is expressible uniquely as a linear combination $g_1\beta_1 + \cdots + g_h\beta_h$ with rational integral coefficients. The numbers β_j are called an integral basis for K, and the discriminant of such a basis is a non-zero rational integer.*

LEMMA 10.7. *If α is an algebraic number in a field K of degree h over the rationals, then the norm $N(\alpha)$, defined as the product of α and its conjugates, satisfies the relation $N(\alpha\beta) = N(\alpha) \cdot N(\beta)$. Also $N(\alpha) = 0$ if and only if $\alpha = 0$. If α is an algebraic integer, then $N(\alpha)$ is a rational integer. If α is rational, then $N(\alpha) = \alpha^h$.*

Finally, from complex variable theory we need the concept of entire function, i.e., a function that is analytic in the whole complex plane, and Cauchy's residue theorem. These ideas can be found, for example, in K. Knopp's *Theory of Functions*, vol. I, Dover, p. 112ff. and p. 130.

3. Two lemmas. LEMMA 10.8.   *Consider the m equations in n unknowns*

(10.1)

$$a_{k1}x_1 + a_{k2}x_2 + \cdots + a_{kn}x_n = 0, \qquad k = 1, 2, \cdots, m,$$

with rational integral coefficients a_{ij}, and with $0 < m < n$. Let the positive integer A be an upper bound of the absolute values of all coefficients; thus $A \geqq |a_{ij}|$ for all i and j. Then there is a non-trivial solution $x_1, x_2, \cdots, x_n$ in rational integers of equations (10.1) such that

$$|x_j| < 1 + (nA)^{m/(n-m)}, \qquad j = 1, 2, \cdots, n.$$

Proof. Write y_k for $a_{k1}x_1 + \cdots + a_{kn}x_n$ so that to each point $x = (x_1, x_2, \cdots, x_n)$ there corresponds a point $y = (y_1, y_2, \cdots, y_m)$. A point such as x is said to be a *lattice point* if its coordinates x_j are rational integers. If x is a lattice point, then the corresponding point y is also a lattice point because the a_{ij} are rational integers. Let q be any positive integer. Let x range over the $(2q + 1)^n$ lattice points inside or on the n-dimensional cube defined by $|x_j| \leqq q$ for all subscripts j. Then the corresponding values of y_k satisfy

$$|y_k| = \left| \sum_{j=1}^{n} a_{kj}x_j \right| \leqq \sum_{j=1}^{n} |a_{kj}| \cdot |x_j| \leqq \sum_{j=1}^{n} Aq = nAq.$$

Thus, as x ranges over the $(2q + 1)^n$ lattice points as indicated, the corresponding lattice points y have coordinates y_k which are integers among the $2nAq + 1$

values 0, ± 1, ± 2, $\cdots$, $\pm nAq$. Thus the lattice points y have at most $(2nAq + 1)^m$ possible locations. We want to establish that at least two of these lattice points y coincide, and this we can do by proving that for some q there are more points than there are possible locations: i.e., that

$$(10.2) \qquad (2q + 1)^n > (2nAq + 1)^m.$$

In order to prove this we now specify the integer q: define $2q$ as the unique even integer in the interval of length 2 defined by

$$(10.3) \quad (nA)^{m/(n-m)} - 1 \leqq 2q < (nA)^{m/(n-m)} + 1.$$

The first part of this inequality implies that $(nA)^m \leqq (2q + 1)^{n-m}$, and so we have

$$(2nAq + 1)^m = (nA)^m \left(2q + \frac{1}{nA}\right)^m < (nA)^m (2q + 1)^m$$

$$\leqq (2q + 1)^{n-m}(2q + 1)^m = (2q + 1)^n,$$

which is (10.2). Thus, as x ranges over the $(2q + 1)^n$ lattice points defined by $|x_j| \leqq q$, the corresponding lattice points y are not all distinct. Say that we have identical points y corresponding to $x = (x_1', \cdots, x_n')$ and $x = (x_1'', \cdots, x_n'')$. Then $x = (x_1' - x_1'', \cdots, x_n' - x_n'')$ gives the non-trivial solution of (10.1) stated in the lemma, because, by (10.3),

$$|x_j' - x_j''| \leqq |x_j'| + |x_j''| \leqq q + q < (nA)^{m/(n-m)} + 1.$$

NOTATION. For any element α of an algebraic number field K, let $\| \alpha \|$ denote the maximum of the absolute values of α and its conjugates. By Lemma 10.4 we note that $\| \alpha + \beta \| \leqq \| \alpha \| + \| \beta \|$ and $\| \alpha\beta \| \leqq \| \alpha \| \cdot \| \beta \|$.

LEMMA 10.9.   *Consider the p equations in q unknowns*

(10.4)

$$\alpha_{k1}\xi_1 + \alpha_{k2}\xi_2 + \cdots + \alpha_{kq}\xi_q = 0, \qquad k = 1, 2, \cdots, p,$$

with coefficients α_{ij} which are integers in an algebraic number field K of finite degree. Assume that $0 < p < q$. Let $A \geqq 1$ be an upper bound for the absolute values of the coefficients and their conjugates for K, thus $A \geqq \| \alpha_{ij} \|$ for all i and j. Then there exists a positive constant c depending on the field K but independent of α_{ij}, p, and q, such that the equations (10.4) have a non-trivial solution $\xi_1, \xi_2, \cdots, \xi_q$ in integers of the field K satisfying

$$\| \xi_k \| < c + c(cqA)^{p/(q-p)}, \qquad k = 1, 2, \cdots, p.$$

Proof. Let h be the degree of K over the field of rational numbers, and let $\beta_1, \beta_2, \cdots, \beta_h$ be an integral basis for the field. If α is any integer of K, then by Lemma 10.6 we can express α uniquely as a linear combination of the integral basis,

$$\alpha = g_1\beta_1 + g_2\beta_2 + \cdots + g_h\beta_h,$$

with rational integral coefficients g_j. Denote the conjugates of α for K by $\alpha = \alpha^{(1)}, \alpha^{(2)}, \cdots, \alpha^{(h)}$, and similarly for the β_j. Taking conjugates in the last equation, by Lemma 10.4 we get

$$\alpha^{(i)} = g_1\beta_1^{(i)} + g_2\beta_2^{(i)} + \cdots + g_h\beta_h^{(i)}, \qquad i = 1, 2, \cdots, h.$$

The determinant $|\beta_j^{(i)}|$ is the discriminant of the basis, and it is not zero by Lemma 10.6. Hence we can solve these equations for the g_j as linear combinations of the $\alpha^{(i)}$, with coefficients dependent only on the basis. Taking absolute values throughout these solutions, we can write

$$(10.5) \qquad |g_j| < c_1 \| \alpha \|, \qquad j = 1, 2, \cdots, h,$$

where the positive constant c_1 is dependent on the field K but independent of α. (In fact, since the discriminant of an integral basis is a non-zero rational integer by Lemma 10.6, we can specify a value for c_1: namely h times the maximum of the absolute values of the first minors of the determinant $|\beta_j^{(i)}|$. However, no use will be made of this specific value.)

To get the integers ξ_i which satisfy the equations (10.4), we write them in terms of the integral basis,

$$(10.6) \qquad \xi_i = \sum_{j=1}^{h} x_{ij}\beta_j, \qquad i = 1, 2, \cdots, q.$$

Thus the problem becomes that of finding suitable rational integers x_{ij}. Equations (10.4) can be rewritten as

(10.7)

$$\sum_{i=1}^{q} \alpha_{ki}\xi_i = \sum_{i=1}^{q} \sum_{j=1}^{h} \alpha_{ki}\beta_j x_{ij} = 0, \qquad k = 1, 2, \cdots, p.$$

Now the integers $\alpha_{ki}\beta_j$ can, by Lemma 10.6, be expressed in terms of the integral basis, say

$$(10.8) \quad \alpha_{ki}\beta_j = \sum_{r=1}^{h} m_{kijr}\beta_r,$$

$$k = 1, \cdots, p, \quad i = 1, \cdots, q, \quad j = 1, \cdots h$$

with rational integers m_{kijr}. Hence (10.7) becomes

$$\sum_{i=1}^{q} \sum_{j=1}^{h} \sum_{r=1}^{h} m_{kijr}x_{ij}\beta_r = 0, \qquad k = 1, 2, \cdots, p.$$

Now the β_r are linearly independent over the rational numbers, and so we set the coefficient of each β_r equal to zero,

$$(10.9) \quad \sum_{i=1}^{q} \sum_{j=1}^{h} m_{kijr} x_{ij} = 0,$$

$$k = 1, 2, \cdots, p, \qquad r = 1, 2, \cdots, h.$$

Here we have ph equations in the qh unknowns x_{ij}, and we will apply Lemma 10.8. In order to do this, we need an upper bound on the absolute values of the coefficients m_{kijr}. We apply the result (10.5), with α replaced by $\alpha_{ki}\beta_j$, and g_j replaced by m_{kijr} as in (10.8), whence

$$| m_{kijr} | < c_1 \| \alpha_{ki}\beta_j \| \leqq c_1 \| \alpha_{ki} \| \cdot \| \beta_j \|$$

$$\leqq c_1 A \| \beta_j \| \leqq c_2 A,$$

wherein A is the bound given in the statement of the present Lemma 10.9, and c_2 is a positive constant chosen not only to satisfy $c_2 \geqq c_1 \| \beta_j \|$ for all j but also so that $c_2 A$ is a rational integer. This last requirement makes c_2 dependent on A, but only in a trivial way which does not affect the order of magnitude of c_2: since $A \geqq 1$, the real number c_2 can be chosen in the interval

$$1 + c_1 \cdot \max_j \| \beta_j \| > c_2 \geqq c_1 \cdot \max_j \| \beta_j \|$$

so that $c_2 A$ is a rational integer.

Hence we can apply Lemma 10.8 to equations (10.9) with m, n, and A replaced by ph, qh, and $c_2 A$. Thus there is a non-trivial solution of (10.9) in rational integers x_{ij} satisfying

$$|x_{ij}| < 1 + (qhc_2A)^{ph/(qh-ph)} = 1 + (hc_2qA)^{p/(q-p)}.$$

Substituting these estimates in (10.6), we get a solution of (10.4) which satisfies the inequalities

$$\| \xi_i \| < h \cdot \max_j \| \beta_j \| \{ 1 + (hc_2qA)^{p/(q-p)} \}$$

$$< c + c(cqA)^{p/(q-p)},$$

provided that c is chosen to exceed hc_2 and $h \parallel \beta_j \parallel$ for all j. Thus c depends on the field K, but is independent of α_{ij}, p, and q. Furthermore, since at least one x_{ij} is not zero, it follows from (10.6) that at least one ξ_i in this solution is not zero, because the β_j are linearly independent over the rational numbers.

4. Proof of the Gelfond–Schneider theorem. We assume that there are two algebraic numbers α and β satisfying the hypotheses of Theorem 10.1, but, contrary to the conclusion of that theorem, we assume that α^β is algebraic. We prove that these assumptions lead to a contradiction. Writing γ for $\alpha^\beta = \exp(\beta \log \alpha)$, we define K as a finite algebraic extension field of degree say h over the rational numbers which contains α, β, and γ. We now write a collection of simple definitions and relationships for reference, and then we will explain them:

$$(10.10) \quad m = 2h + 3, \qquad q > 4m^2, \qquad n = q^2/2m,$$

$$t = q^2 = 2mn, \qquad n > q.$$

The first is a definition of m, so that m and h are fixed throughout the discussion. The integer q is chosen larger than $4m^2$, and so that q^2 is a multiple of $2m$, the integer n being the quotient of these. Then t is defined as q^2, and the last relation, $n > q$, follows from the others. These specifications on q, n, t will be supplemented by one further requirement later: namely that these integers be sufficiently large to exceed certain constants. These constants will be denoted by c, c_1, c_2, c_3, $\cdots$, all of them independent of n, q, t.

Define ρ_1, ρ_2, $\cdots$, ρ_t as the numbers

$$(10.11)$$

$$(r + k\beta) \log \alpha \quad \text{for} \quad r = 1, 2, \cdots, q, \qquad k = 1, 2, \cdots, q.$$

There is no need to specify exactly which of these is ρ_1, which ρ_2, etc.   Define the entire function

$$(10.12) \qquad F(z) = \sum_{j=1}^{t} \eta_j \exp{(z\rho_j)},$$

where the η_j are algebraic integers in the field K to be specified presently.

We use Lemma 10.5 to get a rational integer c_1 so that $c_1\alpha$, $c_1\beta$, and $c_1\gamma$ are algebraic integers.   Then consider the mn equations in the $2mn$ unknowns η_j,

$$(10.13) \qquad c_1^{n+2mq}(\log\alpha)^{-a}F^{(a)}(b) = 0,$$

$$a = 0, 1, \cdots, n-1, \qquad b = 1, 2, \cdots, m.$$

We will apply Lemma 10.9 to these equations, but in order to do so we must verify that the hypotheses of that lemma are satisfied.   The coefficient of η_j in any of the equations (10.13) is, by (10.11),

$$(10.14) \quad c_1^{n+2mq}(\log\alpha)^{-a}\rho_j^a \exp{(b\rho_j)}$$

$$= c_1^{n+2mq}(r+k\beta)^a \exp{\{b(r+k\beta)\log\alpha\}}$$

$$= c_1^{n+2mq}(r+k\beta)^a\alpha^{rb}\gamma^{kb}.$$

This relation establishes that the coefficients of η_j in (10.13) are algebraic integers in K.   To see this, note that the last expression in (10.14) is a polynomial in α, β, γ of degree $a + rb + kb$.   But the maxima of a, b, r, k are $n-1$, m, q, q, respectively, so that $a + rb + kb \leq n - 1 + 2mq$.   Thus the factor c_1^{n+2mq} guarantees that (10.14) is an algebraic integer.

To find a bound for the coefficients (10.14) and their conjugates, we observe that

$$(10.15) \quad \| r + k\beta \| \leq \| r \| + \| k \| \cdot \| \beta \|$$

$$\leq q + q\| \beta \| = q\{1 + \| \beta \|\}.$$

Define c_2 as the maximum of $\|\alpha\|$, $\|\gamma\|$, $1 + \|\beta\|$, and we can say that the absolute value of any coefficient (10.14) or of any conjugate thereof is bounded by

$$c_1^{n+2mq}(qc_2)^n c_2^{2mq} = (c_1 c_2)^n \{(c_1 c_2)^{2m}\}^q (\sqrt{2m})^n n^{n/2},$$

by (10.10). Define c_3 as $(c_1 c_2)^{2m+1}\sqrt{2m}$, and this bound can then be replaced by $c_3^n n^{n/2}$ since $q < n$. Note that c_3, like c_1, c_2, and m, is independent of n.

We can now apply Lemma 10.9 to (10.13) with A replaced by $c_3^n n^{n/2}$, and we conclude that these equations have a non-trivial solution η_j such that, for every j,

$$\|\eta_j\| < c + c\{c(2mn)c_3^n n^{n/2}\}^{mn/(2mn-mn)}$$
$$= c + 2c^2 mn c_3^n n^{n/2} < 3c^2 mn c_3^n n^{n/2}.$$

The constant c is dependent on the field K but not on n. Now $2^n > n > q > m$; so we can replace mn by 4^n in the last inequality, and combine all constants to write

$$(10.16) \qquad \|\eta_j\| < c_4^n n^{n/2},$$

where c_4 is independent of n. This non-trivial solution η_j in integers of the field K of equations (10.13) is used in (10.12), so that $F(z)$ is now completely defined.

LEMMA 10.10. *There exist integers $p \geqq n$ and B in the range $1 \leqq B \leqq m$ such that $F^{(a)}(b) = 0$ for $a = 0, 1, \cdots, p - 1$ and $b = 1, 2, \cdots, m$ and $F^{(p)}(B) \neq 0$.*

Proof. If such an integer p exists, it certainly satisfies $p \geqq n$ because of (10.13). It will suffice to prove that $F^{(a)}(1)$ does not vanish for all the values $a = 0, 1, 2, 3, \cdots, t - 1$. Assume that $F^{(a)}(1)$ does vanish for these values, so that by (10.12)

$$\sum_{i=1}^{t} \eta_j \rho_j^a \exp(\rho_j) = 0, \qquad 0 \leqq a \leqq t - 1.$$

But the η_j are not all zero, so we obtain a vanishing determinant,

$$0 = \det |\rho_j^a \exp(\rho_j)| = \det |\rho_j^a| \cdot \prod_j \exp(\rho_j), \quad 0 = \det |\rho_j^a|.$$

We get the latter equation because $\exp(\rho_j) \neq 0$. By Lemma 10.3 the vanishing of this Vandermonde determinant implies that two of the ρ's are equal, say $\rho_j = \rho_k$. By (10.11) and the fact that $\log \alpha \neq 0$ by hypothesis, it follows that β is rational. But this contradicts the hypotheses of Theorem 10.1, and so the lemma is proved.

Next we use Lemma 10.10 to define the non-zero value

(10.17)

$$\zeta = (\log \alpha)^{-p} F^{(p)}(B) = \sum_{j=1}^{t} \eta_j (\log \alpha)^{-p} \rho_j^p \exp(B\rho_j)$$

$$= \sum_{j=1}^{t} \eta_j (r + k\beta)^p \alpha^{Br} \gamma^{Bk},$$

the latter forms stemming from (10.12) and (10.11); the values of r and k in the last sum depend on j, as in (10.11), but the exact relationship will not be needed in what follows.

LEMMA 10.11.   *There exists a positive constant C, independent of n and p, such that*

$$|N(\zeta)| \geq C^{-p}.$$

Proof.   Recall that the η_j are integers in K, and that the rational integer c_1 was chosen so that $c_1\alpha$, $c_1\beta$, $c_1\gamma$ are integers in K. Hence $c_1^{p+2mq}\zeta$ is an algebraic integer in K, by (10.17) and the facts that q is the maximum value of both r and k in (10.11), and m is the maximum value of B by Lemma 10.10.   Now $q < n \leq p$ so that

$$c_1^{p+2mq} < \{c_1^{1+2m}\}^p = c_5^p,$$

where c_5 like c_1 and m, is independent of p and n. Also $c_5^p \zeta$ is an algebraic integer, so by Lemma 10.7 we have

$$1 \leq |N(c_5^p \zeta)| = |N(c_5^p)N(\zeta)| = c_5^{ph}|N(\zeta)|,$$
$$|N(\zeta)| \geq (c_5^h)^{-p}.$$

This establishes the lemma since h is also independent of n and p.

LEMMA 10.12.   *There exists a positive constant c, independent of n and p, such that $\| \zeta \| < c^p p^p$.*

Proof.   By (10.17) we have

$$\| \zeta \| \leq t \cdot \max_j \{ \| \eta_j \| \cdot \| r + k\beta \|^p \cdot \| \alpha \|^{Br} \cdot \| \gamma \|^{Bk} \},$$

where again the values of r and k are dependent on j, as in (10.11), but the precise relation is of no consequence. Now $q < n \leq p$, and $t = 2mn < 2^n$ for n sufficiently large. Furthermore we can replace r, k, B by their maxima q, q, m from (10.11) and Lemma (10.16) to write

$$\| \zeta \| < 2^n c_4^n n^{n/2} \{ q + q \| \beta \| \}^p \{ \| \alpha \|^m \cdot \| \gamma \|^m \}^q$$
$$\leq \{ 2c_4(1 + \| \beta \|) \| \alpha \|^m \cdot \| \gamma \|^m \}^p n^{n/2} q^p.$$

Also by (10.10) we have

$$q^p = (\sqrt{2m})^p n^{p/2} \leq (\sqrt{2m})^p p^{p/2} \quad \text{and} \quad n^{n/2} \leq p^{p/2}.$$

Applying these to the previous inequality we get

$$\| \zeta \| < \{ 2c_4(1 + \| \beta \|) \| \alpha \|^m \cdot \| \gamma \|^m \sqrt{2m} \}^p p^p = c^p p^p.$$

By Lemma 10.10 the entire function $F(z)$ defined in (10.12) has zeros of order at least p at the points $z = 1$, $2, \cdots, m$. Hence $S(z)$ defined as follows is also an entire function,

$$(10.18) \quad S(z) = p! F(z) \prod_{b=1}^{m} (z - b)^{-p} \prod_{\substack{b=1 \\ b \neq B}}^{m} (B - b)^p.$$

Expanding $F(z)$ in a Taylor series of powers of $z - B$,

$$F(z) = \frac{(z - B)^p F^{(p)}(B)}{p!} + \frac{(z - B)^{p+1} F^{(p+1)}(B)}{(p + 1)!} + \cdots,$$

substituting this in (10.18), and then setting $z = B$, we get

$$(10.19) \quad S(B) = F^{(p)}(B), \qquad \zeta = (\log \alpha)^{-p} S(B),$$

the latter from (10.17).   By Cauchy's residue theorem we obtain

$$(10.20) \qquad S(B) = \frac{1}{2\pi i} \int_C \frac{S(z)}{z - B}\, dz,$$

for any simple closed curve C around the point $z = B$. We shall take C to be the circle $|z| = p/q$. This encloses the point $z = B$ because, by (10.10),

$$(10.21) \qquad \frac{p}{q} > \frac{p}{2q} \geqq \frac{n}{2q} = \frac{q}{4m} > m \geqq B.$$

We now get an upper bound on ζ by use of (10.19) and (10.20).   If u is any complex number, then $|\exp u| \leqq \exp |u|$.   We use (10.11) to write the following inequality for all z on the circle $|z| = p/q$,

$$|\exp (z\rho_j)| \leqq \exp (|z\rho_j|)$$

$$\leqq \exp \left\{ \frac{p}{q} (q + q|\beta|) \cdot |\log \alpha| \right\} = c_6^p,$$

$$c_6 = \exp \{(1 + |\beta|) \cdot |\log \alpha|\}.$$

The constant c_6, like all others in this analysis, is independent of n and p.   Hence by (10.12) and (10.16) we have, for $|z| = p/q$,

$$(10.22) \quad |F(z)| \leqq t c_4^n n^{n/2} c_6^p < (2 c_4 c_6)^p n^{n/2} \leqq c_7^p p^{p/2},$$

wherein we have used $t = 2mn < 2^n \leqq 2^p$ for n sufficiently large. Next, for any $b = 1, 2, \cdots, m$, we use (10.21) to obtain

$$(10.23) \quad |z - b| \geqq |z| - |b| \geqq \frac{p}{q} - m \geqq \frac{p}{2q},$$

$$|z - b|^{-p} \leqq \left(\frac{2q}{p}\right)^p.$$

We now apply (10.22) and (10.23) to (10.18) and conclude that, on the circle $|z| = p/q$,

$$|S(z)| < p! c_7^p p^{p/2} \left(\frac{2q}{p}\right)^{mp} \prod_{\substack{b=1 \\ b \neq B}}^{m} |B - b|^p$$

$$= \{c_7 2^m (2m)^{m/2} \prod_{\substack{b=1 \\ b \neq B}}^{m} |B - b|\}^p p! p^{p/2} \left(\frac{\sqrt{n}}{p}\right)^{mp}$$

$$= c_8^p p! p^{p/2} \left(\frac{\sqrt{n}}{p}\right)^{mp}.$$

Now $p! < p^p$, and $\sqrt{n}/p \leqq 1/\sqrt{p}$ since $n \leqq p$; so we have

$$(10.24) \quad |S(z)| < c_8^p p^{p(3-m)/2},$$

for all z on the circle $|z| = p/q$. Finally we turn to (10.19) and (10.20), and conclude that

$$|\zeta| \leqq |\log \alpha|^{-p} \cdot |S(B)| = \frac{1}{2\pi} |\log \alpha|^{-p} \cdot \left| \int_C \frac{S(z)}{z - B} dz \right|.$$

The length of the path of integration is $2\pi p/q$, and so we can apply (10.24) and (10.23) to write

$$|\zeta| \; < \; |\log \alpha|^{-p} \cdot \frac{p}{q} \cdot c_8^p p^{p(3-m)/2} \cdot \frac{2q}{p}$$

$$< \; \{2c_8|\log \alpha|^{-1}\}^p p^{p(3-m)/2}$$

$$= \; c_9^p p^{p(3-m)/2}.$$

With this estimate for $|\zeta|$, and that of Lemma 10.12 for its conjugates, we write, by (10.10),

$$|N(\zeta)| \; < \; c_9^p p^{p(3-m)/2}(c^p p^p)^{h-1} \; = \; (c_9 c^{h-1})^p p^{-p} \; = \; c_0^p p^{-p},$$

where $c_0 = c_9 c^{h-1}$. This and Lemma 10.11 imply that

$$c_0^p p^{-p} > C^{-p}, \qquad Cc_0 > p,$$

for some positive constants independent of n and p. But this is a contradiction, because $p \geqq n$, and we can choose n arbitrarily large.

Notes on Chapter 10

The special case of Theorem 10.1 for any imaginary quadratic irrational β was established by A. O. Gelfond, *Compt. Rend. Acad. Sci. Paris*, **189** (1929), 1224–1226. The original papers establishing Theorem 10.1 are: A. O. Gelfond, *Doklady Akad. Nauk S.S.S.R.*, **2** (1934), 1–6; Th. Schneider, *J. reine angew. Math.*, **172** (1935), 65–69. The American Mathematical Society has provided an English translation (Translation Number 65) of an advanced expository paper by A. O. Gelfond, *The approximation of algebraic numbers by algebraic numbers and the theory of transcendental numbers, Uspehi Mat. Nauk (N.S.)*, **4**, no. 4 (32), 19–49 (1949). There is an exposition of Gelfond's proof by E. Hille, *Amer. Math. Monthly*, **49** (1942), 654–661.

The proof of Theorem 10.1 given here is based on a simplification of Gelfond's proof by C. L. Siegel, *Transcendental Numbers*, Princeton, pp. 80–83.

Although the methods of Chapters 9 and 10 establish the transcendence of wide classes of numbers, there are many unsolved problems here. For example it is not known whether Euler's constant is irrational, let alone transcendental.

Siegel's monograph cited above is the most complete statement of our present knowledge of transcendental numbers, especially those numbers that arise from solutions of linear differential equations. Finally, we cite some work of a different sort, wherein certain classes of continued fractions are shown to be transcendental: C. L. Siegel, *Abhandl. preuss. Akad. Wiss., Phys.-math. Kl.*, no. 1 (1929), 29; G. C. Webber, *Bull. Amer. Math. Soc.*, **50** (1944), 736–740.

LIST OF NOTATION

(h, k) The g.c.d., or greatest common divisor, of the integers h and k. The same notation is used for the interval from h to k on the real line, where h and k are any real numbers.

$[x]$ The greatest integer less than or equal to x, i.e., the unique integer n satisfying the inequality $n \leqq x < n + 1$.

(x) The fractional part of x; thus $(x) = x - [x]$.

$\phi(n)$ Euler's ϕ function.

J_p The finite field having the p elements $0, 1, 2, \cdots, p - 1$, with addition and multiplication defined modulo the prime number p.

$J_p[x]$ The set of all polynomials in x with coefficients in J_p.

$F_n(x)$ The nth cyclotomic polynomial, i.e., the monic polynomial whose zeros are the primitive nth roots of unity.

R The field of rational numbers.

$[K:F]$ The degree of the field K over the field F, it being presumed that K contains F.

$F(\alpha)$ The field obtained by extending F by the adjunction of α. $F(\alpha)$ consists of the set of all rational functions $f(\alpha)/g(\alpha)$, where f and g are polynomials with coefficients in F, and $g(\alpha) \neq 0$.

$[x_0, x_1, \cdots, x_n]$ A finite continued fraction, whose value can be defined recursively as $[x_0, x_1, \cdots, x_{n-2}, x_{n-1} + 1/x_n]$.

$[x_0, x_1, \cdots]$ An infinite continued fraction, with value $\lim [x_1, x_2, \cdots, x_n]$.

$\subset$ Is contained in.

$\in$ Belongs to, is a member of.

151

$\cup$	The union of; $A \cup B$ is the set of elements x such that x is a member of at least one of the sets A, B.
$d \mid n$	d is a divisor of n.
$d \nmid n$	d is not a divisor of n.
$\exp x$	Same as e^x, the exponential function.
max	The maximum of.
$\| \alpha \|$	The maximum of the absolute values of the algebraic number α and its conjugates.

GLOSSARY

The purpose of this glossary is to supplement, rather than replace, the index. Many of the terms here are not defined elsewhere in this monograph.

algebraic integer. An algebraic number of a special type, namely one that satisfies some equation $x^n + a_1 x^{n-1} + \cdots + a_n = 0$ with rational integral coefficients. The minimal polynomial of an algebraic integer is also monic with integral coefficients.

algebraic number. Any complex number that satisfies some equation of the form $x^n + a_1 x^{n-1} + \cdots + a_n = 0$ with rational coefficients.

almost all. A set S contains almost all real numbers in an interval I if the real numbers (or points) of I not belonging to S constitute a set of measure zero.

Archimedean property. If α and β are any two positive real numbers, then there exists a positive integer n such that $n\alpha > \beta$.

basis of a field. See *finite extension field.*

closed interval. See *interval.*

convergent. See *n-th convergent.*

countable set. A set that can be put into one-to-one correspondence with the natural numbers 1, 2, 3, $\cdots$. See the notes on Chapter 1 for a variation of this definition that is sometimes used.

cyclotomic polynomial. The nth cyclotomic polynomial $F_n(x)$ is the unique monic polynomial whose zeros are the nth primitive roots of unity.

degree of a field. See *finite extension field.*

degree of an algebraic number. The same as the degree of the minimal polynomial of the algebraic number.

denumerable set. One that can be put into one-to-one correspondence with the natural numbers 1, 2, 3, $\cdots$.

division algorithm. Given rational integers a and $b > 0$, one obtains integers q and r such that $a = bq + r$ with $0 \leqq r < b$.

Eisenstein irreducibility criterion. Let p be a prime and $f(x) = a_0 x^n + a_1 x^{n-1} + \cdots + a_n$ a polynomial with integral coefficients such that $p \nmid a_0$, $p^2 \nmid a_n$, $p|a_i$ for $i = 1, 2, \cdots, n$. Then $f(x)$ is irreducible over the rational numbers.

elementary symmetric polynomials. The elementary symmetric polynomials (or functions) of $x_1, x_2, \cdots, x_n$ are the polynomials $\sigma_1, \sigma_2, \cdots, \sigma_n$ defined by the identity

$$\prod_{i=1}^{n} (y - x_i) = y^n - \sigma_1 y^{n-1} + \sigma_2 y^{n-2} + \cdots + (-1)^n \sigma_n.$$

Euler ϕ function. For any positive integer n, $\phi(n)$ is the number of positive integers $j \leq n$ such that the g.c.d. $(j, n) = 1$. Its value is $\phi(n) = n(1 - p_1^{-1})(1 - p_2^{-1}) \cdots (1 - p_r^{-1})$, where the distinct prime factors of n are $p_1, p_2, \cdots, p_r$.

everywhere dense. A set S of real numbers is everywhere dense in an interval I if, given any two numbers α and β in I, say with $\alpha < \beta$, there is a number s in S such that $\alpha < s < \beta$.

field. A system S of elements a, b, c, etc., in which a sum $a + b$ and a product ab are defined for any pair of elements, with the following properties satisfied by all elements: (1) $a + b$ and ab are in S; (2) $a + (b + c) = (a + b) + c$ and $a(bc) = (ab)c$; (3) there exist elements 0 and 1 in S such that $0 + a = a + 0 = a$ and $1 \cdot a = a \cdot 1 = a$; (4) to any element a there correspond elements $-a$ and a^{-1} (only the former in case $a = 0$) such that $(-a) + a = a + (-a) = 0$ and $a^{-1} \cdot a = a \cdot a^{-1} = 1$; (5) $a + b = b + a$ and $ab = ba$; (6) $a(b + c) = ab + ac$. Examples of fields are: the rational numbers; the real numbers; the complex numbers; the finite field denoted by J_p consisting of the numbers $0, 1, 2, \cdots, p - 1$ where p is a prime number, addition and multiplication being defined modulo p.

finite extension field. A field K is said to be a finite extension of a field F if K contains F and if there exists a finite set of elements in K, say $k_1, k_2, \cdots, k_n$, such that every element in K is expressible as a linear combination of these elements, $\Sigma a_i k_i$, with coefficients a_i in F. If no fewer than n elements of K can serve thus, then n is the *degree* of K over F, and the elements $k_1, k_2, \cdots, k_n$ are said to form a basis for K over F. Stated otherwise, the degree of K over F is the maximum number of elements of K that are linearly independent over F.

finite field. A field with a finite number of elements.

fundamental theorem of arithmetic. Any positive integer except 1 can be written as a product of prime factors which is unique apart from the order of the factors.

fundamental theorem on symmetric polynomials. Any symmetric polynomial $f(x_1, x_2, \cdots, x_n)$ with coefficients in a field F is expressible as a polynomial $p(\sigma_1, \sigma_2, \cdots, \sigma_n)$ in the elementary symmetric polynomials with coefficients in F. If the coefficients of f are rational integers, so are the coefficients of p.

Gauss's lemma. If a polynomial $f(x)$ with rational integral coefficients can be factored into two polynomials, $f(x) = g(x)\, h(x)$, with rational coefficients, then there is a non-zero rational number r such that $r\, g(x)$ and $r^{-1}\, h(x)$ have rational integral coefficients.

g.c.d., greatest common divisor. The g.c.d. of two integers a and b (not both zero) is the largest positive integer that divides both a and b; it is denoted (a, b).

half-open interval. See *interval.*

homogeneous polynomial. The polynomial $f(x_1, x_2, \cdots, x_n)$ is *homogeneous* of degree m if $f(\lambda x_1, \lambda x_2, \cdots, \lambda x_n) = \lambda^m f(x_1, x_2, \cdots, x_n)$.

interval. If α and β are any two real numbers, say with $\alpha < \beta$, the real numbers or points x satisfying $\alpha < x < \beta$ form an *open* interval, $\alpha \leqq x \leqq \beta$ a *closed* interval, and $\alpha < x \leqq \beta$ or $\alpha \leqq x < \beta$ a *half-open* interval.

irrational number. A real number that is not rational.

irreducible polynomial. One that is not reducible.

lattice point. In n-dimensional space, any point $(x_1, x_2, \cdots, x_n)$ having integral coordinates.

linearly dependent. The values $\alpha_1, \alpha_2, \cdots, \alpha_n$ are linearly dependent over a field F if there exist elements $c_1, c_2, \cdots, c_n$ in F, not all zero, such that $\sum_{i=1}^{n} c_i \alpha_i = 0$. Otherwise they are *linearly independent.*

measure zero. A set of real numbers or points is said to have measure zero if it is possible to cover the points of the set with a collection of intervals of arbitrarily small total length.

minimal polynomial of an algebraic number. The unique monic polynomial of least degree with rational coefficients which has the algebraic number as a zero.

monic polynomial. One having 1 as the coefficient of the term of highest degree.

n-th convergent. The nth convergent to the continued fraction $[x_0, x_1, x_2, \cdots]$ is the finite continued fraction $[x_0, x_1, \cdots, x_n]$.

open interval. See *interval.*

partial quotient. See *simple continued fraction.*

primitive n-th root of unity. An nth root of unity whose powers give all the nth roots of unity. Such a number is expressible as $e^{2\pi i k/n}$

$= \cos{(2\pi k/n)} + i \sin{(2\pi k/n)}$ for some integer k relatively prime to n.

rational number. One that can be written as the quotient of two integers; thus h/k with $k \neq 0$.

reducible polynomial. A polynomial $f(x)$ with coefficients in a field F is reducible over F if it can be factored non-trivially into two polynomials $g(x)$ and $h(x)$ with coefficients in F, $f(x) = g(x)\,h(x)$. By "non-trivially" we mean that each of $g(x)$ and $h(x)$ has degree at least one.

simple continued fraction. The continued fraction $[x_0, x_1, x_2, \cdots]$ is simple if each *partial quotient* x_i is an integer, positive except perhaps when $i = 0$.

symmetric polynomial. $f(x_1, x_2, \cdots, x_n)$ is symmetric in $x_1, x_2, \cdots, x_n$ if any permutation of the x_j leaves f invariant.

transcendental number. A complex number that is not algebraic.

triangle inequality. Any two complex numbers α and β satisfy

$$|\alpha + \beta| \leqq |\alpha| + |\beta| \quad \text{and} \quad |\alpha - \beta| \geqq |\alpha| - |\beta|.$$

REFERENCE BOOKS

G. Birkhoff and S. MacLane, *A Survey of Modern Algebra*, Macmillan, 2nd edition, 1953.

G. Chrystal, *Algebra*, vol. II, London, Adam and Charles Black, 2nd edition, 1900.

H. Davenport, *The Higher Arithmetic*, Hutchinson's University Library, 1952.

G. H. Hardy and E. M. Wright, *The Theory of Numbers*, Oxford, 3rd edition, 1954.

E. W. Hobson, *Squaring the Circle*, Cambridge, 1913; reprinted by Chelsea, New York, 1953.

E. Kamke, *Theory of Sets*, Dover, 1950.

J. F. Koksma, Diophantische Approximationen, *Ergebnisse der Mathematik*, Band IV, Heft 4, Berlin, Springer, 1936; reprinted by Chelsea.

E. Landau, *Foundations of Analysis*, Chelsea, 1951.

T. Nagell, *Number Theory*, John Wiley, 1951.

O. Ore, *Number Theory and Its History*, McGraw-Hill, 1948.

O. Perron, *Irrationalzahlen*, Berlin, de Gruyter, 1910; reprinted by Chelsea, New York, 1951.

H. Pollard, *Algebraic Theory of Numbers*, Carus Monograph No. 9, John Wiley, 1950.

C. L. Siegel, *Transcendental Numbers*, Annals of Mathematics Studies No. 16, Princeton, 1949.

R. L. Wilder, *Foundations of Mathematics*, John Wiley, 1952.

INDEX OF TOPICS

159

INDEX OF NAMES